Mass Customized Manufacturing

Theoretical Concepts and Practical Approaches

edited by

Vladimir Modrak

CRC Press
Taylor & Francis Group
Boca Raton London New York

CRC Press is an imprint of the
Taylor & Francis Group, an **informa** business

CRC Press
Taylor & Francis Group
6000 Broken Sound Parkway NW, Suite 300
Boca Raton, FL 33487-2742

Printed on acid-free paper by Ashford Colour Press Ltd.
Version Date: 20161020

International Standard Book Number-13: 978-1-4987-5545-0 (paperback)

Library of Congress Cataloging-in-Publication Data

Names: Modrâak, Vladimir, 1957- editor.
Title: Mass customized manufacturing : theoretical concepts and practical approaches / edited by Vladimir Modrak.
Description: New York : Routledge, 2017. | Includes bibliographical references and index.
Identifiers: LCCN 2016026064 (print) | LCCN 2016038648 (ebook) | ISBN 9781315398983 (alk. paper) | ISBN 9781498755450
Subjects: LCSH: Mass customization. | Manufacturing processes. | Industrial management.
Classification: LCC TS183 .M369 2017 (print) | LCC TS183 (ebook) | DDC 670--dc23
LC record available at https://lccn.loc.gov/2016026064

Visit the Taylor & Francis Web site at
http://www.taylorandfrancis.com

and the CRC Press Web site at
http://www.crcpress.com

FSC
www.fsc.org
MIX
Paper from responsible sources
FSC® C011748

Contents

Section III: Management and sustainability of mass customization

Preface

Research interest in mass customization clearly shows that this business phenomenon is on the rise. Bearing in mind the extensive amount of literature in this area as well as scientific publications, one would think that there is not much to say on the subject. And this question occurred to me when I considered whether to start the preparation for this publication. Later, after sorting through all the existing literature, I came to the conclusion that the literature dedicated to the manufacturing perspective of mass customization is not excessive at all. This fact has encouraged me to embark on this journey and to build a wider team of contributors. Some of the invited authors from this research domain perceived the issue as sufficiently explored and elaborated upon, and for this reason they did not express interest in participating in this project. Those who saw the gap in the literature supported my intention and sent me proposals for their intended contributions. Subsequently, the proposed topics for this book were sorted and arranged into three sections and the project moved to its realization phase. It was quite challenging during the editing process to harmonize different research approaches and attitudes into one common body of work. The book you are holding in your hands is the result of these efforts.

The content of the book can be summarily characterized as follows:

The first section presents recent trends and success factors in mass customization. In the introductory chapter, the concept of mass customization and its success factors are briefly outlined. Subsequently, recent trends in the design of mass customization production systems in connection with the *Industry 4.0* concept are presented. The concluding chapter of the first section offers a comprehensive view on the need for a functioning and effective mass customization information system that encompasses all activities involved in the customization process.

Section 2 focuses on complexity drivers, as they are known, whose products and processes in terms of mass customization are becoming more complex. It brings higher uncertainty that may negatively affect the firm's performance results. The first chapter focuses on the role of complexity and its impact on the human operator within assembly-oriented mass customization manufacturing. The second chapter provides a structural framework to model alternative assembly supply chain structures, which are later analyzed using selected structural complexity indicators. Subsequently, the authors focus on a combinatorial-based method to quantify product configurations and variations representing product variety arising in mass customized manufacturing. A method to assess alternative mass customized system designs using newly developed measures is then described. The last chapter of this section presents a method to determine the more suitable degree of customization for a specific product design platform. Its practicality is demonstrated through a case application.

Section 3 mainly provides an overview of product configuration management and its different approaches. The first chapter presents a metamodel of an interactive product configurator to assist the customer during the definition of a customized product. Web-based sales configurators are further discussed in the state-of-the-art analysis for footwear configurators and benchmarked with configurators for the fashion industry. Another chapter investigates the implications of product configurator applications based on a survey and brings answers to questions; for example, how product configurator applications affect companies' business activities. The last two chapters are dedicated to sustainability issues in mass customization manufacturing.

The book provides researchers, scholars, and practitioners with conceptual tools and information about how to use new approaches to maintain and expand market share in a competitive environment. It is also hoped that this book will encourage conventional manufacturers to adopt some of the types of mass customization.

Vladimir Modrak
Technical University of Kosice, Slovakia

Acknowledgments

I am grateful to all chapter authors who contributed to this edited book. I thank my wife Reny for standing beside me throughout its completion. My thanks go also to Dr. Slavomir Bednar, for his continuous and helpful technical assistance on this project, and to my doctoral student, Ms. Zuzana Soltysova, who helped me produce the figures and graphics.

Editor

Vladimir Modrak is full professor of manufacturing technology at Faculty of Manufacturing Technologies and currently deputy rector at Technical University of Kosice. His research interests include manufacturing logistics, cellular manufacturing, lean manufacturing, mass customization, and other related disciplines. He has lectured as a visiting professor at University of Perugia (Italy), Technical University of Applied Sciences Wildau (Germany), and University of Czestochowa (Poland), and has held seminars at the Keyworth Institute of the University of Leeds (UK) and University of Salerno (Italy). Professor Modrak is an editorial board member for several international journals and also serves as ad hoc reviewer for reputable journals such as *International Journal of Production Research, Journal of Cleaner Production, Applied Mathematics and Computation, Entropy, Journal of intelligent and Robotic systems, Journal of Engineering Business Management* and others. He is the leading editor and coauthor of the books, *Operations Management Research* and *Cellular Manufacturing Systems: Innovative Methods and Approaches* (2012) and *Handbook of Research on Design and Management of Lean Production Systems* (2014). He is a Fellow of the European Academy of Industrial Management (AIM).

Contributors

El-Houssaine Aghezzaf is professor of industrial systems engineering and operations research at Ghent University. He is currently heads the Department of Industrial Systems Engineering and Product Design. His main research interests are in integrated optimization and simulation solutions to the design, planning, scheduling, and control problems arising in manufacturing systems and in logistical and utility networks. He has coauthored more than 120 papers, and published in major operations research and industrial engineering journals and conference proceedings. He is associate editor of the *International Journal of Production Research* and a member of the editorial boards of three other journals. He is a member of the executive committee of the Belgian Society of Operations Research and a member of the International Federation of Automatic Control's Coordinating Committee on Manufacturing and Logistics Systems.

Thomas Aichner is a lecturer in consumer behavior at the Faculty of Economics and Management, Free University of Bozen-Bolzano, Italy. He holds a joint PhD in marketing from the University of Padova (Italy) and the ESCP Europe Business School (Berlin, Germany), with the special mention of Doctor Europaeus. His research is focused on international marketing, country of origin, mass customization, e-commerce, and social media. Thomas is coauthor of the 2011 book *Mass Customization: An Exploration of European Characteristics.*

Zoran Anisic is professor of industrial engineering and management at the Faculty of Technical Sciences, University of Novi Sad, Serbia. His area of interest is product development and management, especially mass customization and product lifecycle management. He is founder and manager of My Product, the Center for Product Development and Management, under which he heads a number of departments, including the Mass Customization and Open Innovation Network (MC-OI Network), the International Conference on Mass Customization and Personalization in Central Europe (MCP-CE 2008–2016), and master's studies in product life cycle management. Of all his activities in cooperation with industry, most notable is his development of different types of configurators for local industry. He has published more than 160 papers in conference proceedings and scientific journals.

Dario Antonelli is associate professor of technology and production systems at the Polytechnic University of Turin, Italy. His main research interests concern the development of strategies and methods for human–robot collaborative working cells and the finite-elements simulation and optimization of metalworking processes. He has been team leader of several EC-funded research projects, including ECHORD-FREE, AMICO, @CARE, and ADIUVARE. He is the director of the degree course in engineering and management at the Polytechnic University of Turin, and is a member of the Scientific Committee of PRO-VE and VINORG.

Slavomir Bednar holds a PhD in manufacturing management and is currently occupying the position of assistant professor at the Technical University of Kosice, Slovakia. His research focuses mainly on mass customization, product/process variety, and complexity management, with a particular emphasis on the assembly supply chain complexity in terms of mass customization productions using graph theory, and several kinds of clustering. He is currently engaged in several industrial projects within his research field for the electronic and automotive industries.

Carlo Brondi is a researcher at the Institute for Industrial Technologies and Automation of the Italian National Research Council (ITIA-CNR), with decades of experience in environmental impact assessment. He has been involved in several European, national, and regional projects concerning environmental impact assessment. His main areas of expertise are the sustainable design of innovative production processes and products, life cycle assessment of advanced materials, the implementation of LCA and LCSA approaches in industrial process simulation, the implementation of LCA for company evaluation performance, LCA modularization in the product design phase, and cradle-to-gate and gate-to-grave data integration within certification procedures (an EPD scheme). Research activities concern different industry areas such as automotive manufacturing, machinery, and equipment, the white goods industry, the electronic sector, the leather and wood industry, the packaging sector, and the fashion industry.

Giulia Bruno holds a doctorate in information and system engineering and is currently a postdoctorate researcher at the Department of Management and Production Engineering of the Polytechnic University of Turin, Italy. Her research activity is focused on knowledge management, data mining, and production system analysis. She is working in the fields of human–robot collaboration and product traceability along the whole supply chain. She was involved in several European projects in the context of product life cycle management, health-care systems, and SME networks. Previous research interests also include the analysis of gene expression data to improve tumor classification and the development of data mining algorithms for clinical analysis.

Davide Collatina is a research fellow at the Institute for Industrial Technologies and Automation of the Italian National Research Council (ITIA-CNR). After an initial formation in the LCA of temporary structures in urban areas, his expertise focused on the environmental performance assessment of innovative technologies from the factory field up to the product chain level. Recent application is devoted to LCA and LCSA implementation in the design and simulation of innovative products facing emerging industrial paradigms. Application areas are closed-loop recycling and remanufacturing, industrial symbiosis within industry districts, eco-effectiveness of new technologies, and product customization.

Rosanna Fornasiero works as a researcher at the Institute of Industrial Technologies and Automation of the Italian National Research Council (ITIA-CNR). She has a degree in economics with a specialization in international business and her current research interests include supply chain management, mass customization, and sustainability. In the last 15 years, she has been involved in several European projects related to these topics under the ICT, NMP, and FOF programs. She is author of more than 60 scientific publications and is a member of the IFIP Working Group 5.7. She is coordinator of the road-mapping group of the National Cluster of Intelligent Factories in Italy.

Cipriano Forza is professor of management and operations management at the University of Padova, Italy, where he serves as coordinator of the PhD course in management engineering and real estate economics. He teaches research methods at the European Institute for Advanced Studies in Management (EIASM) in Brussels, Belgium. He served for six years on the board of the European Operations Management Association (EurOMA). He has been published in various journals, including the *Journal of Operations Management*, the *International Journal of Operations and Productions Management*, the *International Journal of Production Research*, the *International Journal of Production Economics*, *Production Planning and Control*, and *Computers in Industry*. His current research focuses on product variety management, including such topics as mass customization, concurrent product–process–supply chain design, and product configuration.

Chiara Grosso is an experienced researcher in economic sociology with a major in technological changes as social and economic processes. She earned the PhD Europaeus at the International University of Catalonia (Barcelona, Spain), where she carried out teaching and research activities as a junior faculty member at the Department of Communication Science. In 2014, she joined the Department of Management Engineering at the University of Padua as a junior research fellow. Her research activities include computer-mediated communication (CMC), social network analysis (SNA), user experience (UX), customer value management (CVM), business digitalization, and web-based product configuration strategies.

Petri T. Helo is professor of industrial management and logistics systems and head of the Networked Value Systems research group at the Department of Production, University of Vaasa, Finland. His research addresses the management of logistics systems in supply–demand networks and the use of IT in operations. Dr. Helo is also a partner and board member at Wapice Ltd., a software solution provider of sales configurator systems and mass customization solutions.

Hendrik Van Landeghem is full professor in operational excellence at Ghent University in Belgium. He is an expert on the design and optimization of manufacturing and logistics systems and their application within industry. He is also an expert in lean management, specifically on implementation methods. He has written more than 100 articles on these and related subjects and contributed to several books, among them one on best practices in operations management and, recently, three volumes on lean implementation methods and flow optimization for SMEs. He is associate editor for the *Engineering Management Journal*. He is a fellow and board member of the European Academy of Industrial Management (AIM) and a fellow of the World Academy of Productivity Science.

Dominik T. Matt is full professor for manufacturing technology and systems at the Free University of Bozen-Bolzano, Italy. He studied mechanical engineering at the Technical University of Munich and earned a PhD in industrial engineering at the University of Karlsruhe. In 1999, he entered the Research and Engineering Center (FIZ) of the BMW Group in Munich. In 2004, he was appointed to the post of professor for manufacturing systems and technology at the Polytechnic University of Turin, Italy. In 2008, he accepted the call of the Free University of Bozen-Bolzano to a tenured professorship at the Faculty of Science and Technology. Since 2010, Professor Matt is also head of the Fraunhofer Innovation Engineering Center (IEC) in Bolzano.

Vladimir Modrak is full professor of manufacturing technology at the Faculty of Manufacturing Technologies, Technical University of Kosice (Slovakia). His research interests include manufacturing logistics, cellular manufacturing, lean manufacturing, mass customization, and other related disciplines. Professor Modrak is an editorial board member of several international journals and also serves as ad hoc reviewer for reputable journals such as the *International Journal of Production Research*, the *Journal of Cleaner Production*, *Applied Mathematics and Computation*, *Entropy*, the *Journal of Intelligent and Robotic Systems*, the *Journal of Engineering Business Management*, and others. He is a fellow of the European Academy of Industrial Management (AIM).

Arun N. Nambiar is an associate professor at California State University, Fresno. His main research interests include software systems as they apply to mass customization, lean principles, and production scheduling. He also works in the areas of RFID and data analytics. He is currently working on the applications of big data in manufacturing.

Erwin Rauch is an assistant professor for manufacturing technology and systems at the Free University of Bozen-Bolzano, Italy. He holds a master's degree in mechanical engineering and in business administration from the Free University of Bozen-Bolzano and the Technical University of Munich, respectively. He obtained his PhD in mechanical engineering summa cum laude from the University of Stuttgart. He was awarded the Overall Best Paper Award at ICAD 2013 and received and a Best Track Paper Award at IEOM 2015. His research interests include the application of axiomatic design, the design of changeable and flexible production systems, the design of distributed manufacturing systems, the design of Industry 4.0 applications for production, and production planning in make-to-order (MTO) and engineer-to-order (ETO) manufacturing.

Enrico Sandrin holds a PhD in operations management and an MS in management engineering from the University of Padova, Italy. His research interests concern mass customization, product configuration, organization design, and human resource management. In his research, he has used both quantitative and qualitative methods, especially surveys and case studies, and a variety of statistical data analysis techniques. Before enrolling in the PhD program, he worked as a knowledge engineer, as a buyer, and as a controller in a firm operating in the machinery industry, where he also had the opportunity to develop strategic improvement projects supported by leading consultancy companies and to act as an in-company trainer. Currently, he is a research associate at the University of Padova, where he also works as a teaching assistant on his topics of expertise.

Nikola Suzic is a lecturer of production system design and decision-making theory at the Department of Industrial Engineering and Management, Faculty of Technical Sciences, University of Novi Sad, Serbia. His main research interest is mass customization, including new product development and production system design for high product variety contexts. The current focus of his research is the application of mass customization in small and medium enterprises (SMEs). Nikola has contributed to three international books, published over 20 papers in international journals and conferences, as well as participated in a number of projects with companies. Since 2006, he has been a member of the Organizing Committee of the International Conference on Mass Customization and Personalization in Central Europe (MCP-CE).

Alessio Trentin is an assistant professor at the University of Padova (Italy), where he received his PhD in operations management in 2006. In 2007–2008, he was a visiting assistant research professor at the Zaragoza Logistics Center (Zaragoza, Spain), a joint research center of MIT (United States) and the Aragona government (Spain). His research interests include mass customization, form postponement, product configuration, build-to-order supply chains, sustainable operations management, and the country-of-origin effect. His work has been published in *Computers in Industry*, the *International Journal of Operations & Production Management*, the *International Journal of Production Economics*, the *International Journal of Production Research*, the *International Journal of Mass Customisation*, the *International Journal of Industrial Engineering and Management*, and the *Journal of Global Marketing*.

Linda L. Zhang is a professor of operations management in the Department of Management at the IESEG School of Management (LEM-CNRS), Lille-Paris, France. She obtained her BEng and PhD degrees in industrial engineering from China in 1998 and Singapore in 2007, respectively. Her research interests include mass customization, the design and management of warehousing systems, health-care service design, and supply chain management. In these areas, she has published many articles in international refereed journals, such as *Decision Support Systems*, *IIE Transactions*, *IEEE Transactions on Engineering Management*, the *European Journal of Operational Research*, the *International Journal of Production Economics*, and so on. Dr. Zhang has had extensive teaching experience in a number of countries, including the Netherlands, France, Singapore, and China. She has taught courses at the undergraduate, graduate, and postgraduate levels.

section one

Trends and success factors in mass customization

chapter one

An introduction to mass customized manufacturing

Vladimir Modrak

Contents

ABSTRACT

In the global business environment, manufacturing companies are facing several important challenges. Among them, nonmarket issues such as environmental and social risk mitigation can be identified. This chapter focuses on the market-oriented challenges associated with the necessity to continuously update product offers in order to serve today's markets and remain competitive. The implementation of an appropriate *mass customization* (MC) strategy provides companies with the most effective way to satisfy individual customers' expectations. This chapter provides a summary of the features and trends of MC and offers new views on the subject.

1.1 About the subject

MC is likely the future trend of business strategy development. In this context, questions arise regarding what characterizes the current trends of MC, how it differs from previous manufacturing strategies, and what will be the future of manufacturing when it takes a global approach. Providing accurate answers is not easy and requires a consideration of at least two aspects of the term. If we comprehend MC as a marketing and manufacturing technique that combines personalized customization and mass production, then we can see manufacturing and marketing perspectives in determining optimal overall strategies for firms.

The first view regards a world of manufacturing that is changing as it follows the world of technology. Technological changes are driven by many factors such as safety and environmental standards, social demands, the diffusion of innovation, and so on. Technology is changing very rapidly and the newest technological developments are reshaping the manufacturing sector in its original form. For example, additive manufacturing (AM), cloud computing, radio frequency identification, fifth-generation (5G) wireless systems, and the Internet of things (IoT) are only a few of the new technologies that are driving a paradigm shift in manufacturing. The umbrella term for this new wave of so-called smart manufacturing is *Industry 4.0*. This concept originates from Germany, but there is a similar view around the world on the future of manufacturing. For instance, the Industrial Internet Consortium in the United States was founded in March 2014 by manufacturing, Internet, information technology (IT), and telecommunications companies, with 212 members as of September 16, 2015. The successive implementation of smart manufacturing capabilities will allow for a faster and better response to customer requirements than ever before. Wide adoption of the IoT into smart manufacturing systems will allow improved flexibility and productivity of the production process and will enable a higher level of MC than is possible today. In this way the manufacturing sector is undergoing a serious transformation that promises other disruptive innovations, including adopting new business models and the production of mass customized products with improved quality and reduced direct costs.

Further development of mass customization from the point of view of the consumer will depend on the willingness of customers to spend time in specifying their preferences and to accept the increased price and delivery time of a customized product. Experiences show that modern consumers desire more and more customized products. At least one of the reasons that consumers prefer custom-made products relates to so-called counterconformity motivation [1]. This kind of motivation is based on the fact that consumers want to be recognized by others as having a particular status in their communities. According to Piller [2], the key element of MC from the consumer perspective is that customers are integrated into value creation as product codesigners by defining, configuring, matching, or modifying an individual solution.

A good starting point in the identification of differences between the current situation of MC and future scenarios is to outline the distinct approaches to MC and its evolutionary development.

1.2 Definitions and approaches to mass customization and its evolutionary development

Since the early 1990s, mass customized manufacturing (MCM) has developed into one of the leading ideas in the production of goods. MC can be understood as the marketing, development, and production of affordable goods or the provision of services with sufficient diversity and adaptation for each customer to customize them to suit his or her needs. In other words, the aim is to give customers what they want, when they want it [3]. Another definition of MC was introduced by Tseng and Jiao [4]. They define MC as the technology and systems producing goods and services based on individual customer requirements.

MCM is a business strategy whereby particular companies expand their sphere of influence by providing attractive opportunities to add value to products and precisely addressing customer requirements. This is also thanks to the fact that consumers are often willing to pay more for customized products than for mass-produced products. On the other hand, MCM is often criticized for the fact that such products do not provide companies comparable profit to traditional customized production. In fact, MC is still a new manufacturing concept that is rarely applied by companies and requires further research and development.

The most recent classification of MC approaches was offered by Kull [5], who divided MCM into two types: configuration oriented and parameterized oriented. If customers can configure selected parts of products individually to their own solution, this is the configuration type. The second one allows customers to change the visual aspects of the product—that is, the shape and/or size of the product's components. According to Gilmore and Pine [6], MC may appear as (1) *adaptive customization*, (2) *cosmetic customization*, (3) *transparent customization*, or (4) *collaborative customization*. When applying adaptive customization, products and services are standardized but have a few customized options. For example, a company offers a package of software designed to run all the activities of small businesses, and the buyer as a final customer can add more accounting functions into the package.

Cosmetic customization means that a company produces products that are standardized, but the market offers the products in different ways to different customers. The main difference between adaptive and cosmetic customization is in the offer of standardized products, where adaptive customization offers only a choice of standardized products, and cosmetic customization offers the customer groups of standardized products. It means that the customer can choose from an apparent variety of offered products. For example, companies offer different sizes of products, different colors, different appearances, and so on.

The transparent form of customization is driven by modularity. Customers as codesigners simply combine different parts or parameters of a predefined set of components according to their needs. Products are in this way customized for customers. For example, car producers offer customers their own car configurators on their web pages, where customer can choose from groups of component options and build the car to his or her specifications.

Collaborative customization is when the producer conducts a dialogue with the individual customer. This form of customization allows a customer to select from groups of standardized component/parameter options and specify undefined features or properties of the product. An example of collaborative customization can be taken from the shoe industry. Customers can customize their own products online and give them specific lines. These simplified descriptions of the four types of customization are graphically depicted in Figure 1.1.

Lampel and Mintzberg [7] have proposed five strategies, from *pure standardization* to *pure customization*, by which business models leading to a customization-oriented production can be categorized. In pure standardization, there is a dominant design targeted to a wide group of customers. This strategy conforms to adaptive customization. The second strategy, classified as *segmented standardization*, targets small groups of customers that have better choices than in the previous strategy. This business model corresponds well with the category of cosmetic customization. The next category is so-called *customized standardization*, where product configurations are designed by customers. This business model is more or less identical to transparent customization. *Tailored customization* as the further development of MC offers a generic product model that is tailored according to the customer's individual needs. This business strategy is more or less similar to the model of collaborative customization. And finally, the highest level of MC is pure customization, where the optimal level of personalization is achieved and the customer can build the product according to his or her individual specifications. These classification frameworks offer important insights into the development of manufacturing policy and practice, but it does not mean that these classification frameworks will fit any original equipment manufacturer.

1.3 Success factors for mass customization

An elementary condition for the application of MC is the consideration and analysis of alternative strategies aimed at increasing competitiveness through innovative product design and customer satisfaction management. A well-known fact is that overall customer satisfaction is higher when the product better matches the customer's ideal preference.

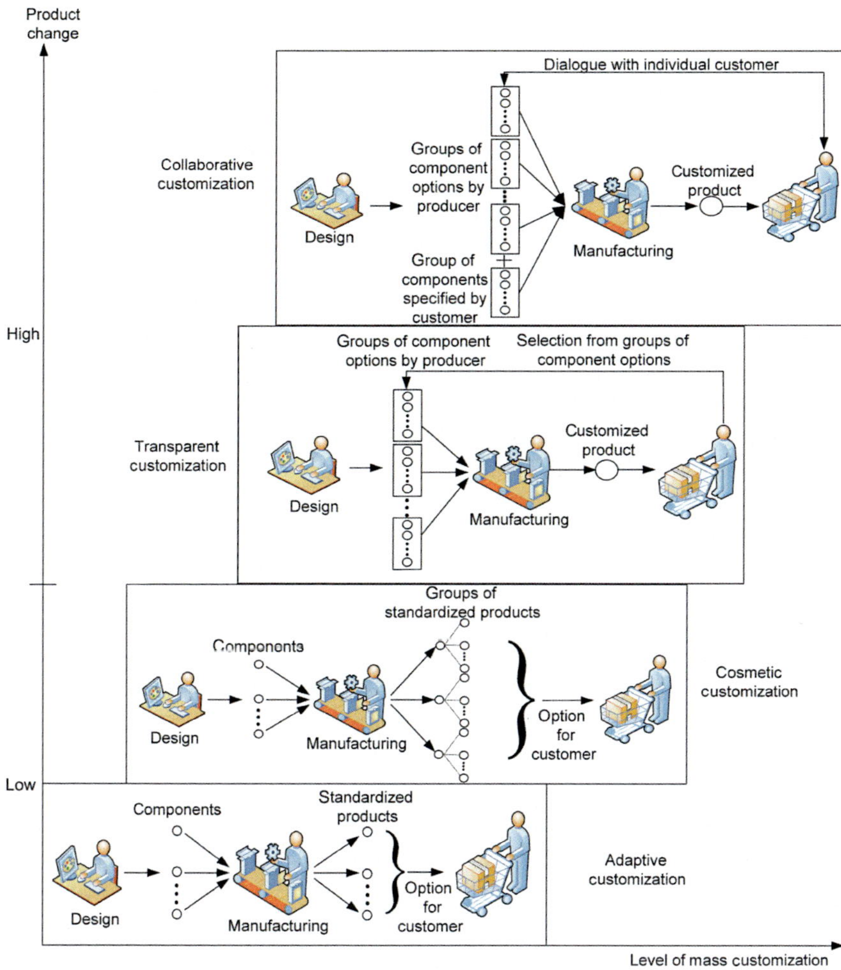

Figure 1.1 Four levels of MC.

Long-standing strategies by which this objective can be achieved are product proliferation and product customization.

A firm pursuing product proliferation will offer many different product types with different features, functions, and so on. When applying this strategy, production and logistics costs can be negatively affected by the number of different products. If increasing numbers of products are provided by a firm on a replenishment basis, then its suppliers have to expand their product lines. Such a situation makes it more difficult to forecast demand and call out for a transition from a *make-and-sell* model to a so-called *sense-and-respond* organization. The first model is focused on production efficiency and the second one is customer satisfaction oriented. Moreover, the sense-and-respond organization business model allows all members of the supply chain to adapt to changing market conditions and to work together seamlessly [8].

Product customization can be defined as producing physical goods that are tailored to a particular customer's requirements [9]. This strategy makes it possible to meet each customer's specific needs more precisely than through product proliferation, although the level of taylorization is limited. According to Zipkin [10], increasing the complexity of MC processes can potentially limit the degree to which customization is beneficial to customers. A reasonable degree of customization depends on several factors, such as the kind of industry a company is part of, the level of manufacturing flexibility, clients' wishes, and so on. It is rather difficult for firms to find optimal rates of customization for an existing or new product due to the wide range of opinions represented by the number of offered product configurations. For this purpose, the generic concept of identifying an optimal degree of product customization can be used (Figure 1.2). This balancing concept adopts Tainter's curve of complexity [11] and ensures that products are neither under-customized nor over-customized.

As outlined previously, there are at least two ways to deliver a higher level of product variety, and MC may not always be the best. Therefore, an early and reliable decision whether MC is the right prescription for a firm or not is a critical step toward achieving sustainable development objectives. To answer this, the following four fundamental factors can be effectively analyzed.

1. *Customers' readiness*. According to Da Silveira et al. [12], "Mass customization encompasses the ability of original equipment

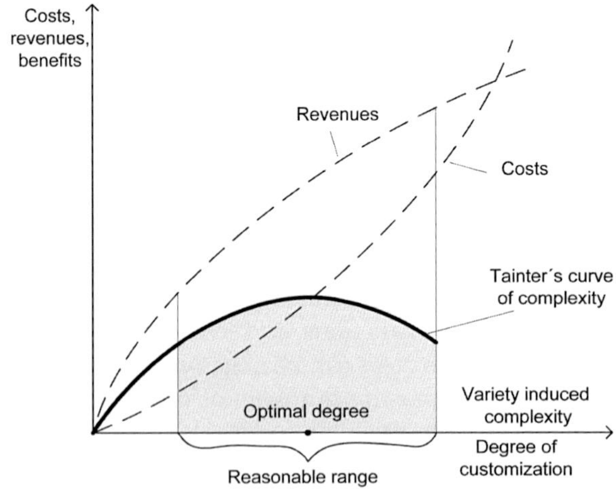

Figure 1.2 Generic concept of identifying the optimal degree of product customization.

manufacturers (OEMs) and their suppliers to provide individually designed products and services to customers in the mass-market economy." Notable attention is paid to the role of OEMs, who sell their products to personal consumers. Guilabert and Naveen [13] argue that knowing how significant customization is for potential consumers as well as how it varies by type of product will help producers to implement one of the customization strategies.

Basically, customers' readiness can occur in explicit or implicit form. Based on this categorization, the following construct of *customers' customization readiness* (CCR) can be outlined. So-called explicit customers' readiness for buying customized products can be seen in daily life. For example, most people prefer customizing their furniture by choosing the style, size, and finish to suit their individual needs, rather than purchasing standard products. In such situations, MC is directed by a customer's specific needs and whims (Figure 1.3a).

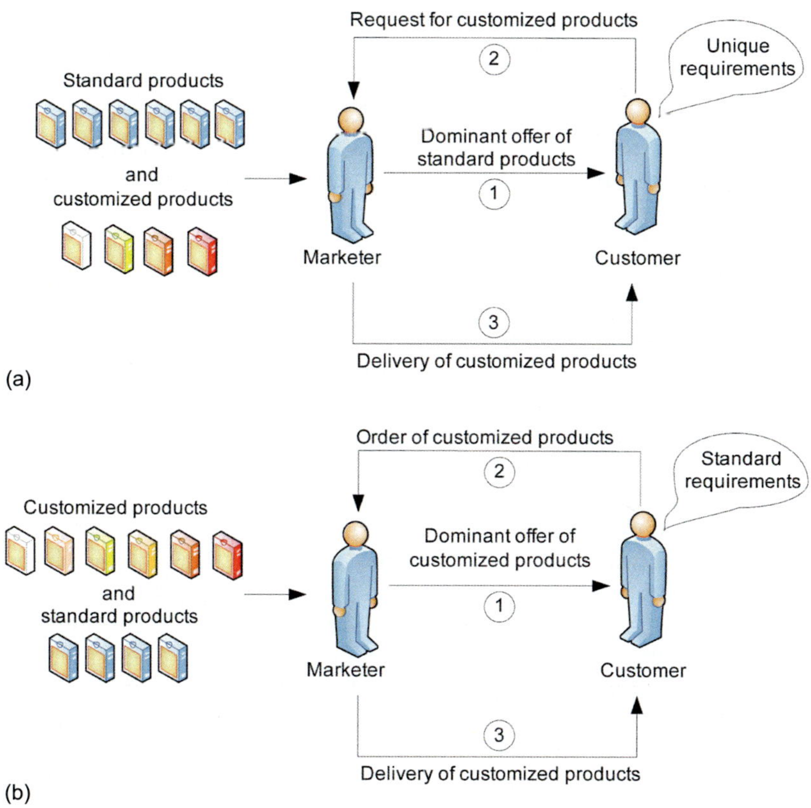

Figure 1.3 Customers' customization readiness construct: (a) explicit form; (b) implicit form..

So-called implicit customers' readiness for buying customized products has to be revealed through interaction between the marketer and customer by offering a full range of customization options. When applying such an approach, one can then say that MC is pushed by marketer's options (Figure 1.3b). Both types of CCR are highly pivotal in paving a way toward implementing an MC strategy. The proposed CCR construct differs from the *customer customization sensitivity* (CCS) construct developed by Hart [14]. The CCS construct is based on two factors: uniqueness of customer needs and customer sacrifices. According to Hart, the level of CCS is directly proportional to the uniqueness of customer needs and/or customer sacrifices.

2. *Type of products.* According to Da Silveira et al. [12], MC will never be possible for all types of products. In this sense, Duray [15] argues that "the production of standardized modules is the key to high volume mass customization." In addition, Tsang and Hu [16] point out that convenient products for MC are those with short life cycles, and Blecker [17] emphasizes that "companies have to offer tailored products while ensuring short delivery times simultaneously." Taken together, the following factors predispose products to MC:
   a. Products can be grouped together into a product family.
   b. Products are designed as modules so that they can be easily assembled into different ones.
   c. Product can be delivered in short times.
3. *Market characteristics.* Some markets offer full customized product varieties, whereas in other markets they are mostly available as discrete product varieties. Obviously, under specific conditions, both types of markets can bring opportunities or indicate threats when a company is intending to implement one of the MC strategies. Cavusoglu et al. [18] identified the following market categories as critical in choosing an optimal customization strategy: *competition, segmentation,* and *cannibalization.*

   The intensity of competition significantly influences customization possibilities and predetermines a choice of the strategic patterns of the players in the market. Pine [19] recommends three ways to shift to MC: incrementally over time, more quickly through business transformation, or by creating a new business. The incremental path toward an MC strategy, apart from other factors, assumes that competing firms that operate in the same reference market are not delivering customized products.

   However, this can be a slow process if competitors are already effectively issuing MC. For companies facing such competition, rapid transition to an MC business strategy helps them remain competitive in the marketplace. Increased need for new products in a highly dynamic market environment can be reflected by transforming businesses to a higher level of customer satisfaction. Pine [19] recommends that transforming business in such a way can be achieved by creating a group of related businesses focused on individual customers' needs.

Segmentation is often the key to developing a competitive edge. Research conducted by Jiang [20] shows that MC is not totally the same as segmenting to one. In this context, it has been highlighted that firms that aim customization at specific consumer segments may not be optimal [18].

The cannibalization effect is understood as the extent to which a product variety reduces a firm's profits from the standardized varieties it produces. Yayla-Küllü et al. [21] argue that the cannibalization effect dominates in a highly competitive market. A useful insight to this effect is provided by Selladurai [22].

4. *Firms' readiness* for MC can be understood as having the right conditions and resources in place to support the transformation process. Knowing that MC means a huge variety of products by combining a large number of product modules, companies that want to follow this path firstly have to analyze their technological, organizational, and managerial capabilities to determine whether they are potentially transformative to this strategy. El Kadiri et al. [23] predict that in an MCM environment, intelligent device technologies using sensors and actuators will dominate. It is expected that the extension of 5G technologies to the ICT sector will facilitate automation, which brings for manufacturing a new impetus to foster MC. Rapid technological progress continues in AM technologies. This paradigmatic change in manufacturing poses significant challenges for enterprises to utilize the technology for MC. For example, laser-sintering machines present an attractive mode of production that has great potential to widely customize products such as bone implants, prostheses, medical devices, and so on. Equally, direct metal laser sintering is widely and effectively used to fabricate metal prototypes. When looking at economic evaluation, cost comparisons between AM technologies show that traditional processes are more economically effective than AM technologies in high output quantities [24,25]. Thomas and Gilbert [26] in this context state that these viewpoints come from analyses of well-structured AM costs, and they add that significant benefits and cost savings in AM may be hidden in the ill-structured costs. An important advantage of AM technologies is design freedom. Reeves et al. [27] point out that due to these design freedoms, assembly operations that were previously required to build a complex component can be reduced. Moreover, AM technologies remove the risk of the long lead time for

the delivery of tooling [28]. When assuming the need for the combination of these trends and technologies, in particular, MC calls for the development of entirely new business organizations.

1.4 *Some current and future trends in mass customization*

The most important enablers of MC are web-based product configuration systems. Taking this fact into account, Blecker et al. [33] developed a customer needs model providing decision makers with insights concerning product variety management problems. The basic idea of this model lies in eliminating product variants that only correspond to subjective needs. The authors emphasize that this principle might be incorporated into online configuration systems.

It is expected that the diffusion of digital manufacturing technologies will trigger a transformation from MC to mass personalization. Kumar [34] argues that IT capabilities will drive MC programs toward the mass personalization strategy. Mass personalization differs from MC in many aspects. While MC assumes stable product architecture and product modules, for mass personalization, possible changes of the basic design architecture and product features are typical [35]. However, wider acceptance of this strategy in individual industries will strongly depend on the availability of attainable digital manufacturing tools and other devices belonging to the *smart manufacturing* concept.

1.5 *Concluding remarks*

Initially, MC was seen as a contradictory approach that could not lead to entrepreneurial success. Despite its conflicting ideas, the existence of MC is a reality thanks especially to the advances realized in the fields of flexible manufacturing and IT. As was predicted in earlier as well as more recent literature [36–39], MC has become an imperative rather than a choice for success and sustainability across business sectors.

References

1. Tepper, K., Bearden, W. O., and Hunter, G. L. (2001). Consumers' need for uniqueness: Scale development and validation. *Journal of Consumer Research*, 28(1), 50–66.
2. Piller, F. T. (2004). Mass customization: Reflections on the state of the concept. *International Journal of Flexible Manufacturing Systems*, 16(4), 313–334.
3. Pine, B. J. (1993). *Mass Customization: The New Frontier in Business Competition*. Brighton, MA: Harvard Business Review Press; 1st ed. (October 1, 1992), p. 368.

4. Tseng, M. M., and Jiao, J. (1998). Design for mass customization. *Annals of the CIRP*, 45, 1.
5. Kull, H. (2015). *Mass Customization: Opportunities, Methods, and Challenges for Manufacturers*. New York: Apress, p. 148.
6. Gilmore, J. H., and Pine, B. J. (1997). The four faces of customization. *Harvard Business Review*, 75(1), 91–101.
7. Lampel, J., and Mintzberg, H. (1996). Customizing customization. *Sloan Management Review*, 38, 21–29.
8. Haeckel, S. H. (1999). *Adaptive Enterprise: Creating and Leading Sense-and-Respond Organizations*. Brighton, MA: Harvard Business School Press.
9. Blecker, T., Friedrich, G., Kaluza, B., Abdelkafi, N., and Kreutler, G. (2005). *Information and Management Systems for Product Customization*, Vol. 7. Berlin: Springer Science+Business Media.
10. Zipkin, P. (2001). The limits of mass customization. *MIT Sloan Management Review*, 42(3), 81–87.
11 Tainter, J. A. (2006). Social complexity and sustainability. *Ecological Complexity*, 3(2), 91–103.
12. Da Silveira, G., Borenstein, D., and Fogliatto, F. S. (2001). Mass customization: Literature review and research directions. *International Journal of Production Economics*, 72(1), 1–13.
13. Guilabert, M. B., and Naveen, D. (2006). Mass customisation and consumer behaviour: The development of a scale to measure customer customisation sensitivity. *International Journal of Mass Customisation*, 1(2–3), 166–175.
14. Hart, C. W. (1995). Mass customization: Conceptual underpinnings, opportunities and limits. *International Journal of Service Industry Management*, 6(2), 36–45.
15 Duray, R. (2002). Mass customization origins: Mass or custom manufacturing? *International Journal of Operations and Production Management*, 22(3), 314–328.
16. Tseng, M. M., and Hu, S. J. (2014). *Mass Customization: CIRP Encyclopedia of Production Engineering*. Berlin: Springer, pp. 836–843.
17. Blecker, T. (2005). *Mass Customization: Concepts-Tools-Realization*. Berlin: GITO Verlag.
18. Cavusoglu, H., and Raghunathan, S. (2007). Selecting a customization strategy under competition: Mass customization, targeted mass customization, and product proliferation. *IEEE Transactions on Engineering Management*, 54(1), 12–28.
19. Pine, B. J. (1999). *Mass Customization: The New Frontier in Business Competition*. Brighton, MA: Harvard Business Press.
20. Jiang, P. (2000). Segment-based mass customization: An exploration of a new conceptual marketing framework. *Internet Research*, 10(3), 215–226.
21. Yayla-Küllü, H. M., Parlaktürk, A. K., and Swaminathan, J. M. (2013). Multi-product quality competition: Impact of resource constraints. *Production and Operations Management*, 22(3), 603–614.
22. Selladurai, R. S. (2004). Mass customization in operations management: Oxymoron or reality? *Omega*, 32(4), 295–300.
23. El Kadiri, S., Grabot, B., Thoben, K. D., Hribernik, K., Emmanouilidis, C., von Cieminski, G., and Kiritsis, D. (2015). Current trends on ICT technologies for enterprise information systems. *Computers in Industry*, 79, 14–33.
24. Ruffo, M., Tuck, C., and Hague, R. (2006). Cost estimation for rapid manufacturing–laser sintering production for low to medium volumes. *Proceedings of the Institution of Mechanical Engineers, Part B: Journal of Engineering Manufacture*, 220(9), 1417–1427.

25. Deradjat, D., and Tim, M. (2015). Implementation of additive manufacturing technologies for mass customisation. *International Association for Management of Technology IAMOT 2015 Conference Proceedings*, 24, 2079–2094.
26. Thomas, D. S., and Gilbert, S. W. (2014). *Costs and Cost Effectiveness of Additive Manufacturing.* Gaithersburg, MD: National Institute of Standards and Technology (NIST), 1176.
27. Reeves, P., Tuck, C., and Hague, R. (2011). Additive manufacturing for mass customization. In *Mass Customization*, Folliatto, F. S., and Da Silveira, G. J. C. (Eds.), pp. 275–289. London: Springer.
28. Hopkinson, N., Hague, R., and Dickens, P., eds. (2006). *Rapid Manufacturing: An Industrial Revolution for the Digital Age.* New York: Wiley.
29. Blecker, T., Abdelkafi, N., Kaluza, B., and Kreutler, G. (2004). A framework for understanding the interdependencies between mass customization and complexity. 1–15. In: *Proceedings of the 2nd International Conference on Business Economics, Management and Marketing*, Athens, Greece: Athens Institute for Education and Research (ATINER) (2004): pp. 291–306.
30. Kumar, A. (2007). From mass customization to mass personalization: A strategic transformation. *International Journal of Flexible Manufacturing Systems*, 19(4), 533–547.
31. Tseng, M. M., Jiao, R. J., and Wang, C. (2010). Design for mass personalization. *CIRP Annals—Manufacturing Technology*, 59(1), 175–178.
32. Piller, F. T., and Tseng, M. M., eds. (2010). *Handbook of Research in Mass Customization and Personalization*, Vol. 1. Singapore: World Scientific.
33. Anderson, D. M. (1997). *Agile Product Development for Mass Customization.* Chicago: Irwin.
34. Kratochvíl, M., and Carson, C. (2005). *Growing Modular: Mass Customization of Complex Products, Services and Software.* Berlin: Springer-Verlag.
35. Kumar, U., Ahmadi, A., Verma, A. K., and Varde, P., eds. *Current Trends in Reliability, Availability, Maintainability and Safety: An Industry Perspective.* Switzerland: Springer International 738 pages, 1st edition, 2016.

chapter two

Designing assembly lines for mass customization production systems

Dominik T. Matt and Erwin Rauch

Contents

ABSTRACT

This chapter focuses on the design of assembly lines and systems for the production of mass customized products. It reviews the state of the art in mass customization (MC) and mass customized manufacturing (MCM) to give an overview of existing research, actual findings, and future trends of manufacturing and assembly systems for this application. A framework for a systematic design of assembly for MC products is described. The framework discusses product design requirements, the design of mass customization assembly (MCA) systems, the requirements for MC shop floor management in assembly, as well as opportunities for MCA through the fourth industrial revolution (*Industry 4.0*). Afterward, a real case study at a medium-sized manufacturer of electromotors shows an example of successful assembly systems design for make-to-order production with high product variety and mass customized products. The experience from the case study shows a concept based on the combination of manual assembly and automated stations as well as the integration of radio frequency identification (RFID) technology in the assembly process. The chapter closes with a brief summary of the guidelines for assembly systems design for MC and the experience from the case study. Further, the conclusions give an outlook to future research activities.

2.1 Introduction

MC as a combination of mass production and customized production is the response from the manufacturing industry to the requirements of an increasingly modern and dynamic world. Society is constantly developing new opportunities and challenges for individuality in consumption. The modern individualist eats his or her own cereal selection ordered online, wears tailor-made shirts, and drives an individually configured vehicle. People's increasing desire for individuality promotes the trend toward MC. In addition to this consumer-driven trend, however, there are often very practical and pragmatic reasons for a request for customized products. The producer who is able to produce a product at low prices and with customer-specific characteristics as quickly as possible has the highest competitive advantage on the market. Simultaneously, new technologies such as advanced web technology and additive manufacturing have opened new possibilities for capturing customer requirements and for producing customer-specific products. This chapter is mainly concerned with the production of customized products, with a special focus on assembly processes and systems. In traditional mass production, assembly lines are often highly automated and designed as single-model lines or lines with a small number of variants. Due to the high degree of automation, costs can be reduced to a minimum, but at the same time flexibility is seriously restricted. As we will see in this chapter, MC combines these highly controversial objectives and allows maximum flexibility, producing products with a reasonable cost structure. This chapter intends to give systems designers support for the design of MCA systems by means of a systematic framework. In addition to product-related requirements for MC, the design of MCA systems and their operation as well as their integration into Industry 4.0 are treated.

2.2 Theoretical background

MC has been identified as a competitive advantage strategy by an increasing number of companies (Da Silveira et al., 2001). The concept of MC was first expounded formally in the book *Future Perfect* by Stanley M. Davis (1989). MC means the production of products that have been customized for the customer at production costs similar to those of mass-produced products (Kaplan and Haenlein, 2006). MC allows customers to select attributes from a set of predefined features in order to design their individualized product, by which they can fulfill their specific needs and take pride in having created a unique result (Hart, 1995; Schreier, 2006; Stoetzel, 2012). Thus, customization integrates customers with the design process (Qiao et al., 2006). The primary focus of the product designer should be on providing value to the end user. To achieve personalization, increasing the value for the end user, different authors suggest a user-centered design approach (Kramer et al., 2000). The end user, in the sense of open innovation and the democratization of design, is more often directly or indirectly involved in the product development process. Therefore, innovation, in the future, will take place not only within the company but can be seen as an interactive process between the company and the market to generate customized products (Reichwald and Piller, 2005).

Numerous authors have published articles about MC; many of them discuss MC from a strategic and economic point of view. Only a few research works investigate the technical aspects of the manufacturing and manufacturing systems design of mass customized products (Koren, 2005; Terkaj et al., 2009; Mourtzis et al., 2013; Bednar and Modrak, 2014; Matt et al., 2015). MC brings radical changes to methods used to operate traditional manufacturing enterprises. It is changing the way customers make purchases, and this has a strong impact on how products are manufactured (Smirnov, 1999). MCM has been gaining recognition as an industrial revolution in the twenty-first century. Customers usually can select options from a predetermined list and request them to be assembled (Qiao et al., 2006). While the manufacturing industry in the past distributed globally standardized products to keep the production cost and complexity low, nowadays the customization of products based on customer-specific needs is becoming more and more important (Matt et al., 2015). Simultaneously, with this development in the direction of an increasing number of individual product variants and product configurators, the requests for production and the complexity of manufacturing systems increased. Manufacturing systems in an MC environment should be able to produce small quantities in a highly flexible way and be rapidly reconfigurable (Qiao et al., 2006; Thirumalai and Sinha, 2011).

In the eighties, the concept of *flexible manufacturing systems* (FMSs) was introduced in response to the need for MC and for greater responsiveness to changes in products, production technologies, and markets (Wiendahl et al., 2007). Traditional FMSs, including the popular Japanese lean production manufacturing system, have not reached the flexibility and adaptability demanded by an MCM system (Koste and Malhotra, 1999; Matt et al., 2014). The concept of *reconfigurable manufacturing systems* (RMSs) emerged later in the nineties in an attempt to achieve changeable functionality and scalable capacity (Koren, 2005). In recent years, the concept of *changeable* and/or *agile manufacturing systems* has developed with the ability of factories or production systems to switch from one product family to another, changing the production capacity accordingly (El Maraghy and Wiendahl, 2009; Wiendahl et al., 2009). Such a manufacturing system fits the requirements for MCM better than a traditional one.

In addition to the basic design of production systems for MC, several scientists were engaged in the design of assembly systems in particular. Assembly is one of the most cost-effective approaches to high product variety, allowing the production of mass customized products nearly at mass production costs (Hu et al., 2011). The aim is a codesign process with an open product architecture (Koren et al., 2013) in combination with on-demand manufacturing systems, enabling user participation in the design, product simulation, manufacturing, supply, and assembly processes that rapidly meet consumer needs and preferences (Tseng and Hu, 2014). The economy of scale is achieved at the component level, while the economy of scope of high variety is achieved in the final assembly using flexible and reconfigurable assembly systems (Hu et al., 2011). Modern approaches for designing assembly lines for MC often show two general alternatives for multiple product models: (1) a multimodel assembly line where products are produced on the same line but in batches for each product model; and (2) mixed-model assembly lines, where the product model variants are sufficiently similar that they can be assembled simultaneously on the same line (Boysen et al., 2007). Many of these approaches can also be found in practice. BMW claims that "every vehicle that rolls off the belt is unique" (Zuh et al., 2008). Such a situation presents enormous difficulties in the design and operation of assembly systems. It has been shown by empirical and simulation studies (MacDuffie et al., 1996; Fisher and Ittner, 1999) that increased product variety has a significant negative impact on the performance of assembly processes. The higher the number of product variants and configurations or overall variety, the more complex difficulties there are in the production design and operational management of assembly systems or assembly supply chains (Modrak et al., 2015).

The latest trend in MC is digitalization in manufacturing, also known by the term *Industry 4.0* or *cyber-physical systems* (CPSs). The large potential of Industry 4.0 will be a key enabler for further developments in MCM (Kull, 2015). Dombrowski et al. (2013) describe in their work a concept for a *cyber-physical assembly system* (CyPAS) with modular CPSs supporting the handling of high-level customization. Liu et al. (2014) propose a concept for an Internet of Things (IoT)-enabled intelligent assembly system, in order to improve the interconnection, perception, efficiency, and intelligence of assembly systems. Intelligent, cognitive, and self-optimizing manufacturing systems are able to learn and thereby perform self-determined changes in production systems (Schmitt et al., 2012). To reach such a level of changeability, it is necessary to equip manufacturing systems with cognitive capabilities in order to take autonomous decisions in even more complex production processes with a high product variety (Zäeh et al., 2009).

2.3 Framework for the design of highly changeable production systems for mass customization assembly

This chapter presents a framework for the design of highly changeable production systems for MCA. Figure 2.1 illustrates four stages of the framework, described later in detail. First of all, product design–related aspects for the design of MC products are shown, putting the focus on design aspects for increasing the *assemblability* of MC products. Then the design of changeable assembly systems is described with an approach called *SMART reconfigurability*. A production system consists not only of the assembly system itself, but also of organizational aspects about operation and shop floor management. Thus, specific shop floor arguments for MCA are explained in the third section. Due to the actual challenge to become Industry 4.0 ready, the last

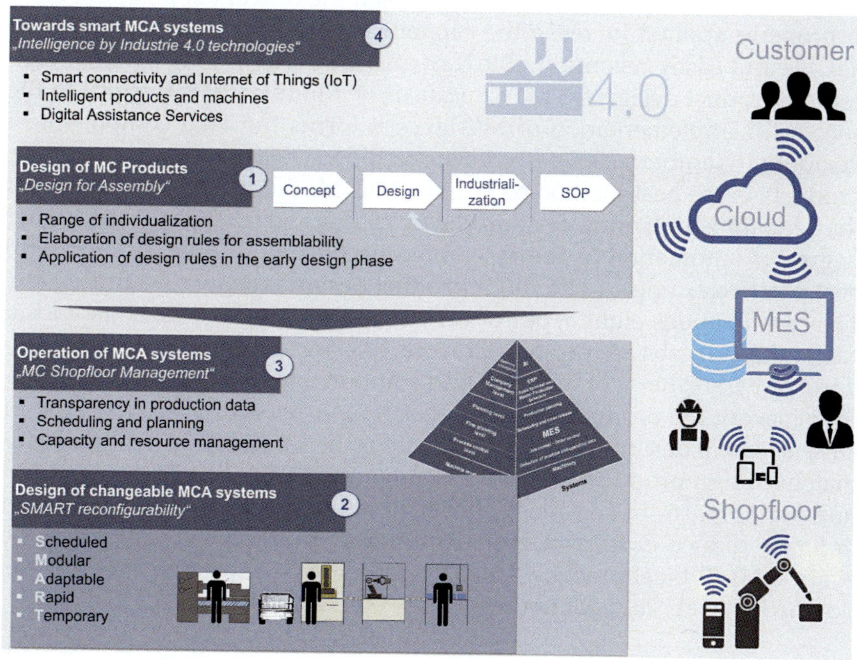

Figure 2.1 Framework for MCA.

section discusses advanced technologies such as RFID, the IoT, and others, with the aim to achieve a smart MCA.

2.3.1 Product design–related aspects for mass customization assembly

It is generally estimated that up to 70%–80% of product life cycle costs are determined during the design phase (Whitney, 1988). Therefore, product designers should take into account what the requirements are to simplify and accelerate the assembly of mass customized products. Assemblability is defined as a combination of the following three aspects in assembly (Lefever and Wood, 1996):

1. The number of operations required to assemble the product
2. The handling of each needed component
3. The complexity of the required operations

A fundamental technique to reduce complexity in the manufacturing and assembly processes of products with high variety is *postponement*. The concept is to design products in such a way that the point of differentiation into multiple product variants is delayed as much as possible within the production sequence. Products to which the concept of postponement is properly applied increase the capability of an assembly system to be flexible and easily reconfigurable (Lee and Billington, 1994). Another concept in product design for MC is *modularity*. Modularity is integral to the successful implementation of MC since it forms the vital component of flexibility (Nambiar, 2009). According to Pine (1993), developing modular products is the best method to achieve MC. Modular design can address the need for a high number of product variants and further allow a higher degree of automation in the assembly line (Salonitis, 2014). A successful best-practice example of modular product design is Scania AB in Sweden. The case includes eight types of cabs with thousands of variants within each type. All cabs are produced on one line and are built up from a standardized assortment of modules and components (Eastman, 2012). A further concept for product design is *design for assembly* (DFA). The goal of DFA is to help designers explicitly consider the assembly process and ultimately design products that are assembled with the minimum required number of parts in the most efficient and economical way possible in order to reduce error and cost (Anisic and Krsmanovic, 2008; Arnette et al., 2014). Typical practical rules for DFA are (Lefever and Wood, 1996; Boothroyd et al., 2010; Jakubowski and Peterka, 2014)

- Reduce the total number of parts.
- Develop a modular design.

- Use standard components.
- Design parts to be multifunctional.
- Minimize assembly directions.
- Minimize the complexity of assembly operations.
- Minimize the handling of parts and products.

An interesting approach for the design of mass customized goods is shown in six steps by Hernandez et al. (2006).

1. Definition of the space of customization
2. Formulation of an objective, which can often be the minimization of cost
3. Identification of modes for managing product variety such as adjustable controls, modular combinations, and dimensional customization that are used to customize the product
4. Determination of the number of hierarchy levels and allocation of the modes for managing product variety to these levels
5. Formulation of a multistage optimization problem
6. Problem-solving with a focus on cost-effectiveness, suitability for small or large variety in product specification, and adaptability of the product line (Figure 2.2)

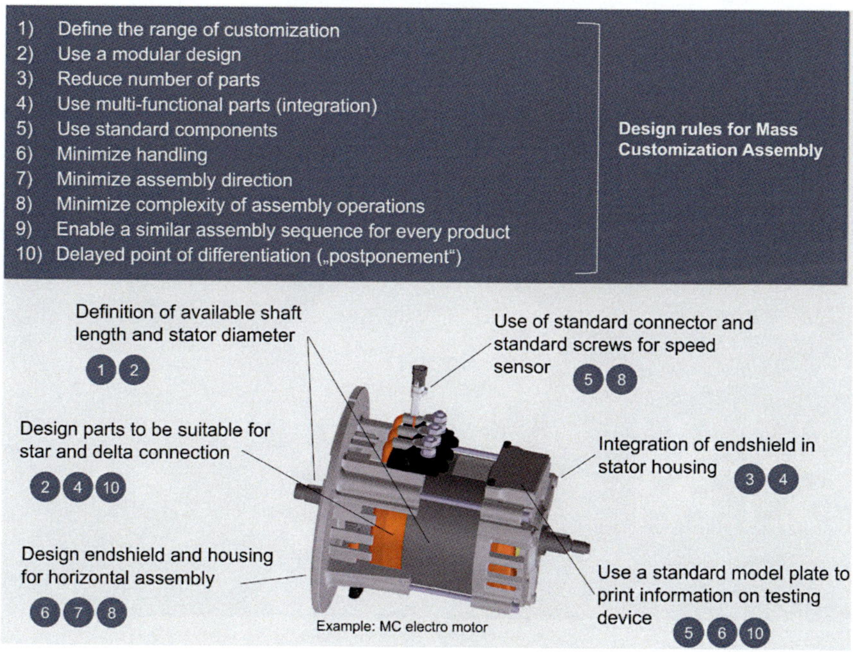

Figure 2.2 Design rules in product design for MCA.

2.3.2 *Design of flexible and reconfigurable mass customization assembly systems*

Based on the concept of flexibility and reconfigurability in manufacturing systems, this section outlines design guidelines for MCA systems. The concept of *SMART reconfigurability* will be explained (Matt et al., 2014), where assembly systems should be designed in ways that can be quickly adaptable to different product variants. The proposed approach includes the principles and concepts already applied in industrial manufacturing to design flexible and changeable production systems and extends them to fulfilling the requirements of MC-oriented manufacturers. These companies have to adapt their manufacturing and assembly systems frequently and temporarily. Sometimes, the requirements for the next reconfiguration are not even known in advance or depend on personalized features of the product. Thus, a manufacturing or assembly system that is able to fulfill these requirements may also be called a SMART system. To obtain a highly flexible and reconfigurable assembly system for an MC environment, five major requirements can be identified (see also Figures 2.3 and 2.4).

- Scheduled
- Modular
- Adaptable
- Rapid
- Temporary

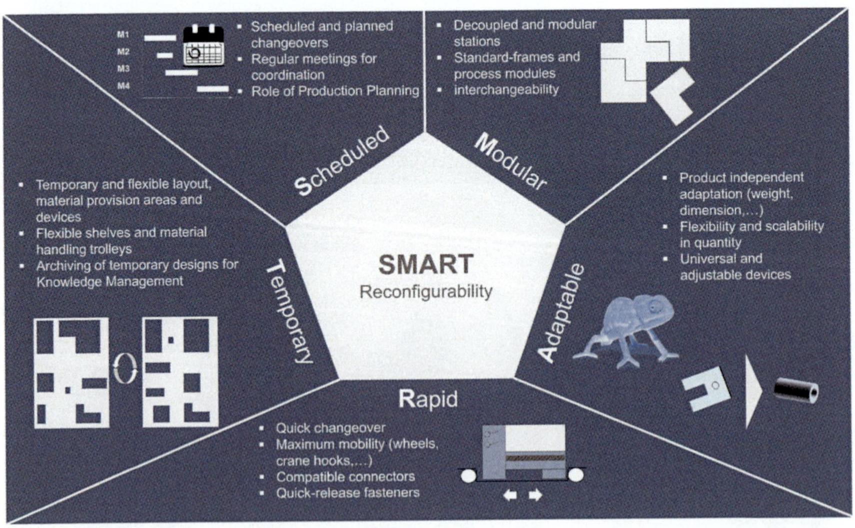

Figure 2.3 SMART reconfigurability approach. (From Matt, D.T., et al., *Enabling Manufacturing Competitiveness and Economic Sustainability*, Springer, 2014.)

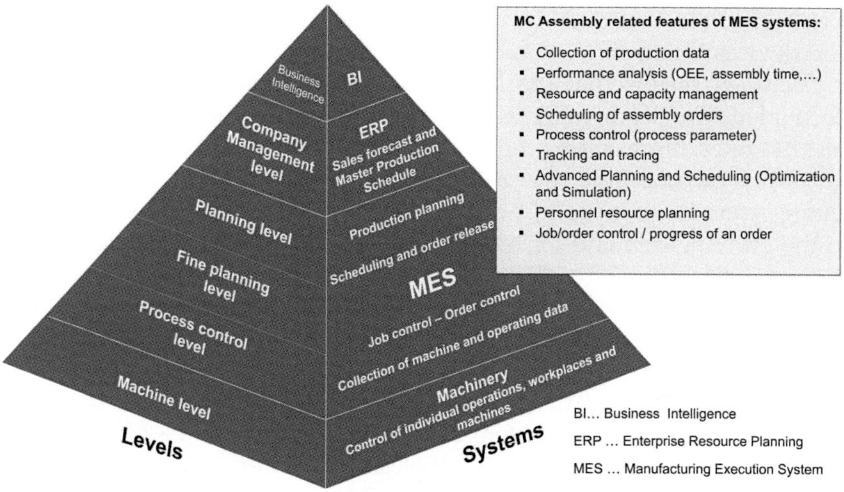

Figure 2.4 Typical elements of a MES for MCA.

In this concept, the term *SMART* stands for

1. Scheduled and planned changeovers
2. Modular and decoupled assembly stations
3. Adaptable stations to a varying number of configurations
4. Rapid ability to change from one configuration to the other
5. Temporary, due to the very limited period of validity of assembly configurations

Scheduled stands for planning and scheduling every major reconfiguration of the assembly system. Often, inefficiencies arise from the inadequate planning of the necessary reconfigurations in assembly. Unlike traditional production, the assembly of a variety of different and personalized product variants requires accurate planning—for example, with the support of digital planning tools such as a *manufacturing execution system* (MES). The information transfer between order processing and production has to be accelerated, so that a suitable planning and scheduling of necessary reconfigurations can be enabled and carried out in time. Therefore, for the implementation of a precise and perfectly working production, planning is a prerequisite for reconfiguration in assembly. Also, transparency in the order sequence and regular coordination meetings between production and production planning are necessary elements. For this purpose, in the case study, the role of the manufacturing system designer is introduced. When the production schedule is defined, the production engineer becomes responsible for defining, organizing, and executing all the necessary changes of the production system to guarantee a material flow–oriented layout and the availability of the required tools and devices for an efficient and ergonomic assembly and material supply.

Modular means that the individual assembly work stations can be decoupled. This leads to increased flexibility in the use and reuse of process modules, representing the key functionality of modularity. Increasing flexibility on one side, the reconfiguration needs solutions for quick changeovers to achieve short-order execution and setup times. In relation to the growing demand for modular systems, the producers of assembly systems or cells have reacted in recent years by developing and offering such systems. On the German market, the first pioneers were producers such as teamtechnik with a system named TEAMOS, ZBV-Automation with a system named CORAcell, as well as Paro offering the first modular assembly systems in 2007 (Drunk, 2007). In the sector of MC, this ability to adapt an assembly system as quickly as possible to a new product or variant is indispensable. Nevertheless, the aforementioned industrial assembly systems are suitable also for MCA.

Adaptable is equivalent to changeable and describes the ability of the assembly system to be reconfigured or adapted to produce a high variety of products. Thus, such an adaptable preassembly requires an extremely high degree of changeability and flexibility. A high degree of adaptability for MC companies includes not only the adaptability of individual machines or assembly workstations, but also the ability for a quick and easy reconfiguration and setup of the entire assembly line before beginning a new variant. This includes the design of universal or variable and adjustable process elements or fixing systems (*external flexibility*), as well as a high amount of *internal flexibility*. Internal flexibility means that the assembly system is able to be modified and reconfigured for the next product variant without any physical change of the layout or of the process elements. A simple example of internal flexibility is the identification of a new product variant through barcode scanning or reading the RFID tag of a new product configuration (operating parameters—e.g., torque for wrenches) and the subsequent automatic setup of the assembly line or cell. Today, flexible programmable industrial robots or assembly devices as well as collaborative robots with an automatic setup function through controlled axes are able to reduce setup times substantially.

Rapid describes the need of the assembly systems to switch as soon as possible to another configuration, reducing waste and increasing efficiency in operation. As we know, setups in the assembly systems restrict the available and value-added time fundamentally and can therefore have a high impact on the cost of mass customized goods. Thus, the rapid adaptation of the assembly system has a direct impact on the success and profitability of MC-oriented companies. The system designer has to be very considered in implementing maximum mobility (wheels, roles, crane hooks, integrated apertures for forks, etc.) and compatibility (standard connectors), and rapid disassembly and assembly (e.g., using quick-release fasteners instead of bolts and nuts). It is important to not only optimize the hardware but also to train the employees in the methods of lean and quick changeover (e.g., the single minute exchange of dies [SMED] method) to minimize the downtime.

Temporary characterizes the challenge that an MCA system is only valid for a very limited period and has to be reconfigured very often. While the reconfiguration of assembly in many other industries, such as the electronics and consumer goods industries, is often limited only to switching to another similar product variant, in MC the assembly system can change substantially, depending on the defined range of customization (see Section 2.3.1). Thus, assembly processes, layouts, material provisions, and tools are only temporary in nature and are subject to constant change. Therefore, a system designer should avoid providing material through rigid systems and transfer points, using more likely mobile or flexible shelving and trolleys. This represents a challenge for the system designer as well as for all involved support departments, such as quality management. Processes and tools for monitoring the process stability and quality of products must be constantly adapted in parallel with the assembly system. To achieve this, many companies try to install screens at the workplace, integrating digital-quality management systems that can be easily changed at the moment of the reconfiguration of the assembly line.

2.3.3 *Operation and shop floor management of mass customization assembly systems*

In the last decade, shop floor management has been mainly optimized by methods from lean production, achieving significant savings and productivity gains. With the introduction of production monitoring software, all production data can be processed in real time to provide the necessary information for operative production management. Shop floor management comprises not only the control of orders in real time, but also the optimization of inventory in the production, increasing the overall equipment effectiveness (OEE), reliable maintenance and assistance, as well as just-in-time material supply through organizational measures and IT systems.

MCM does not only rely on advances in manufacturing technology. The organization of the production process is an important prerequisite to customizing products efficiently (Blecker and Friedrich, 2007). MC production companies must fight with shop floor uncertainty and complexity caused by a wide variety of product components (Zhong et al., 2013). Planning and scheduling therefore become difficult at the shop floor level in the case of make-to-order industries. Due to many different product variants, mixed-model production, and the need for just-in-sequence, the material supply and order management at assembly lines is very complex. Many components of MC products are one of a kind and need to be supplied at the right time of the assembly process. While production planning and order processing are used mainly to deliver products on time, the traditional aim of shop floor management is to maximize machine utilization and productivity (Huang et al., 2009). These challenges in the operation and management of the shop floor in MC has encouraged many firms to introduce enterprise resource systems (ERS) and later also MESs. Such manufacturing IT systems deployed on the shop floor enhance manufacturing precision and process flexibility, thus supporting the development of MC capability. In addition, they enable fast and efficient manufacturing operations by integrated access to production-related data (Xiaosong Peng et al., 2011). The modern MES mainly focuses on shop floor operations such as scheduling, execution, and control, informing shop floor supervisors in terms of manufacturing progress, equipment status, material delivery, and consumption (Blanc et al., 2008). The ISA 95 standard shows a common framework and terminology of MES processes and allows the classification and comparison of MES software systems (Cottyn et al., 2011). Upon receiving information on the MC product, the MES can look up suitable and available stations at which operations can be performed. It decides the operations to be scheduled on the machines, choosing the best fit by means of utilization rate, energy efficiency, and delivery time. After planning the production, the MES sends the processing instructions to the selected stations and hence starts the production (Keddis et al., 2013).

Figure 2.4 illustrates typical features of an MES for MC production and assembly. The figure shows the different system levels in the IT landscape of manufacturing firms. Business intelligence (BI) is a strategic tool for transforming general data into meaningful information for top management decisions. Enterprise resource planning (ERP) is needed for general data management and for the execution of sales forecasts and rough capacity planning. Between ERP and the physical machine level, we can find the MES level for planning, execution, and process/order control. Figure 2.4 shows the MCA-related features of a typical MES to support operations and shop floor management. These features range from the simple collection of production and machine data to performance analysis, resource management, advanced planning and scheduling functions, order release and control, as well as tracking and tracing.

2.3.4 *Smart mass customization assembly systems: Opportunities through the fourth industrial revolution*

The first industrial revolution was characterized by the mechanization of production, while the second industrial revolution introduced mass production, followed by the third digital industrial revolution: the use of electronics and IT as well as automation. Industry 4.0 refers to the fourth industrial revolution, with a technological evolution from embedded systems to CPSs and the introduction of the IoT to production. The main objectives of Industry 4.0 are the reduction of complexity in industrial production processes by the use of "intelligent" structures and CPSs enhancing decentralized intelligence, and networking with the interaction of the real world and the virtual environment. New digital information, communications, and web technologies seem to act as enablers to shift from *centralized* to *decentralized* production, where the product communicates with the manufacturing system and the linked software applications and databases. In the context of Industry 4.0, new information and communications technologies act as enablers of smart, autonomous, and self-learning factories. With the technological opportunities of Industry 4.0, shop floor management should become mobile, digitally visualized, and even more smart and intelligent at the same time. Industry 4.0 will only work if machines can communicate between each other and material flows are tracked by RFID or similar technologies throughout large sections of the industry. Self-controlling systems communicate via the Internet among themselves and with the human operators in assembly and manufacturing systems (Brettel et al., 2014).

Figure 2.5 shows an overview of possible applications of Industry 4.0 technologies in the field of MCA. New technologies will have a major

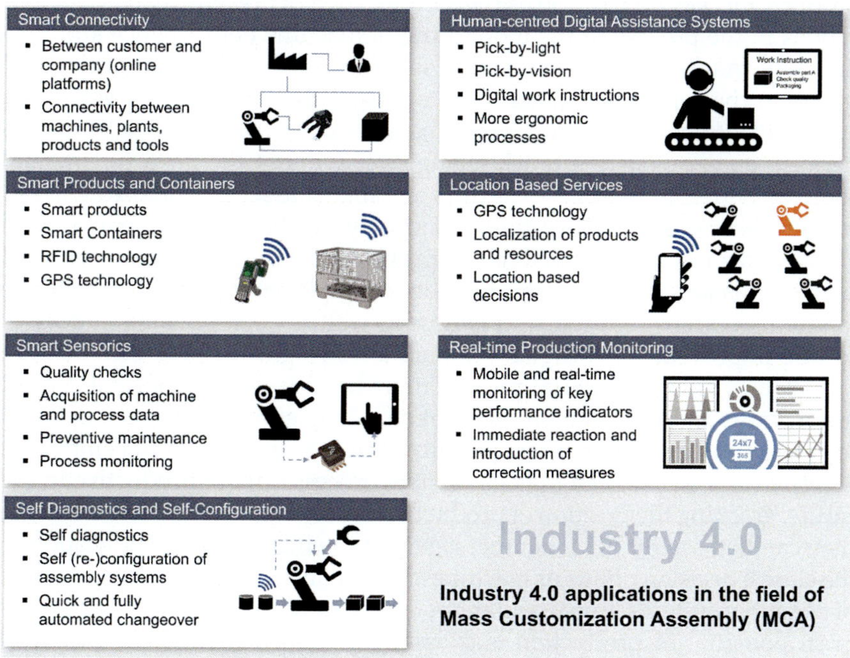

Figure 2.5 Industry 4.0 applications in the field of MCA.

impact on how jobs or product data are generated. Through the creation of online platforms for product design and product configuration, items are individually created and forwarded directly to the producing companies. *Smart connectivity* here is not only the keyword for the exchange of data between customer and company, but also for transparency in the communication between machines, plants, products, warehouses, and tools.

Smart products also make up almost a prerequisite for the use of many Industry 4.0 applications in production and assembly. Through RFID-enabled real-time manufacturing execution systems, assembly objects can be tracked and traced systematically, collecting real-time production data. This allows for the identification of disturbances and inefficiency at the assembly line much earlier than with traditional instruments (Zhong et al., 2013). RFID with an agent-based workflow management system can be focused to improve the shop floor's configurability and reusability during the manufacturing process (Zhang et al., 2014). If products cannot be equipped with RFID it is possible in many cases to create smart containers or boxes equipped with RFID technology. Another application of advanced technologies in production and assembly are *smart sensorics*. Through wireless data acquisition in the assembly process or in the machine, data can be checked, used, and tracked any time and anywhere in the world. Other applications for smart sensorics can be found in the detection of incorrect products for rejection as well as the identification of preventive maintenance actions in the process or the monitoring of machine performance and productivity. As just mentioned, smart sensorics enable self-diagnostics and thus also the self-configuration of intelligent assembly systems. When the system identifies a product change, the assembly system should be able to perform autonomously a quick changeover without the need for manual setup activities. In most cases, a fully automated assembly process will not be suitable or possible. Therefore, in Industry 4.0, the human operator plays a major role toward productivity improvement. Through human-centered *digital assistance systems*, the productivity of employees can be increased, the failure rate can be reduced, and fatigue as well as nonergonomic processes can be improved. Such digital support technologies include pick-by-light or pick-by-vision in material provision or the visualization of digital work instructions and information on a screen. Using employee badges or RFID tags, this information could be visualized in the language of the employee. *Location-based services* can make assembly processes more productive in the future. Tracking the position of production supervisors or maintenance staff, they can be brought via push message and intelligent alarm notification to the closest decision points or machines, thus reducing unplanned downtime of assembly systems. In particular, the transparency of real-time data shows great potential in production and assembly. Information about assembly performance at the end of the shift is only a retrograde point of view and prevents the supervisor from introducing correction measures. Through mobile real-time production monitoring, every level of the company—from top management to employees on the line—has transparency about the status of the assembly system at all times. This allows problems to be tackled immediately and thus productivity and OEE can be increased.

2.4 Industrial case study

The concepts and design approaches for MCA have been applied to a project with a medium-sized manufacturer of electric motors. The company has about 400 employees and is one of the world's leading manufacturers of electrical drive systems for mobile vehicles. The company's strength lies not in producing cheap standard motors for standard applications, but in the development and production of customized drives with individual customer requirements. The company currently produces approximately 900 different variants of AC and DC motors. Of these, only 20 types are A-articles with a large demand of up to 10 units/day, while the remaining motors have small-to-medium order quantities. Approximately 60% of the motors sold are mass customized products or small series with a quantity of 1–10 units/year. Based on this analysis, the company wanted to develop a new concept for the assembly, providing a separation of the production into a *high runners* area and an MC area. Figure 2.6 summarizes the most

Figure 2.6 Improvements in MCA at a manufacturer of electric motors.

important improvements in the MC production area. In what follows, these improvements are explained more in detail.

In the first step, the project team analyzed the products through an ABC analysis and a detailed analysis of every single assembly step. The aim was to determine the similarity of products and variants in their assembly processes and to cluster them into product families. Based on the results of the investigation and the determined product families, the articles were taken together to develop a new optimized assembly concept for every product family. In this first phase of the project, various meetings were also held with the engineering department to define for each product family the range of individualization. The following main criteria were defined: the number of wire turns, the size limits of the motor, the length limits of the iron plates, and the individualization limits in terms of customer-dependent optional parts and components.

In the next step, the integration of RFID tags on the product was discussed with the engineering department. By applying RFID labels, the product can be equipped with intelligence: production data can be saved during the production process, the product can be traced throughout the entire production process, and at the same time, internal logistic processes can be synchronized with the real-time order status. In addition to the logistics optimization in the production process, the logistic process in shipping and in the preparation of transport documents can be realized more efficiently. The work effort for counting the motors manually, booking them out of the warehouse system, and then assigning them to the transport documents can be eliminated or reduced to a minimum. With the integration of RFID gates in the shipping area, the forklift truck driver can drive through the gate without waiting and attach the automatically printed pallet label; the transport documents will then be automatically generated. At the same time, the error rate is reduced, as the system performs an automatic comparison between the goods on the truck and the ordered goods.

Every assembly station was equipped with a digital touchscreen monitor. When the product passed, related information, such as drawings, work instructions, or notes (error messages and causes), was visualized on the screen, identifying the product with the RFID tag.

In the assembly line, the work piece carrier was standardized for every product family. If standardization was not possible, the carrier was designed such that employees can switch easily and rapidly between the individual sizes by folding work piece holders. In the past, the required pressing processes were executed at different assembly stations with up to three consecutive presses. In the new assembly system, those processes were aggregated into a fully automated universal pressing device. The new universal pressing device identifies autonomously the product by reading the RFID tag, starting a setup process when the product is changing. Using electric drives in the device, distances and positions for pressing are adapted fully automatically (self-configuration). Thus, the time for changeover could be reduced to a minimum without any manual tool change, as was needed in the "old" assembly lines.

In the new assembly process, several smart sensors and cameras were integrated for quality checks and the control of process stability. For example, the automated process control system performs an automated presence control of locking rings and seals. With the help of smart sensorics, the system measures process parameters as well as environmental parameters (temperature, humidity, etc.) to investigate long-term statistics about the influence of environmental changes on process stability and quality. In addition, data requested by the customer, such as the force–distance relation as well as final quality testing results, is stored on the RFID tag.

With the use of warehouse management software and an electronic kanban system, material provision for assembly could be improved. Purchased parts and components can now be supplied just in time and, if needed, just in sequence, checking the actual progress of the product through defined scanning points in the production and assembly process.

Data is now collected at every single station by an MES. The MES supports not only the collection of machine and process data (utilized for initiating continuous optimization measures on the shop floor level) but also planning functions working with a real-time manufacturing execution system. Due to the high transparency in production and machine data, the productivity can be visualized in real time on monitors as well as in a smartphone app for supervisors and top management.

The new assembly concept has already been implemented successfully in part and is showing promising results. A full implementation of all the aforementioned and elaborated measures will take place in 2016. Thus, the full impact and effects of the new assembly concept will be measurable in 2017.

2.5 Conclusion and outlook

Due to the increasing individualization of products and increasingly complex customer requirements, the trend of MC will rise even farther in the future. Similarly, as shown in the industrial example, companies have to adapt their production accordingly to manage the realization of customer-specific products with a batch size of one or the smallest lot sizes, without a disadvantageous increase in costs. Especially in assembly, small batch sizes and a wide range of variants lead to a high degree of complexity in the design of the assembly system and in its operation. MC causes not only difficulties in machinery design but also a certain cognitive complexity for operators in assembly environments. As indicated in the chapter, MCA systems have to be designed in a changeable and flexible way. A framework for the design of such assembly systems was introduced. It was pointed out that a lot of the problems in engineering could be solved by taking the DFA approach. The approach of SMART reconfigurability gives system designers a set of guidelines for the planning and realization of modern assembly systems for MC. In addition, the framework also shows the importance of operational shop floor management in MCA systems. At this level, manufacturing companies are introducing increasingly specific IT tools such as MESs. Finally, new opportunities for advanced technologies were discussed. For the purposes of Industry 4.0 and the IoT, assembly systems should be designed to be smart and intelligent. In modern assembly systems, important information can be sent between machines and humans in real time and visualized continuously and anywhere, with the scope to improve efficiency.

Future research should examine the possibilities of 4.0 technologies, to support the implementation of MC in assembly as well as in the entire production system. Further investigation in the field of modern and intelligent production and assembly systems thus represents a significant "booster" for MC in industry.

References

Anisic, Z., and Krsmanovic, C. (2008). Assembly initiated production as a prerequisite for mass customization and effective manufacturing. *Strojniški vestnik*, 54(9), 607–618.

Arnette, A. N., Brewer, B. L., and Choal, T. (2014). Design for sustainability (DFS): The intersection of supply chain and environment. *Journal of Cleaner Production*, 83, 374–390.

Bednar, S., and Modrak, V. (2014). Mass customization and its impact on assembly process' complexity. *International Journal for Quality Research*, 8(3), 417–430.

Blanc, P., Demongodin, I., and Castagna, P. (2008). A holonic approach for manufacturing execution system design: An industrial application. *Engineering Applications of Artificial Intelligence*, 21(3), 315–330.

Blecker, T., and Friedrich, G. (2007). Guest editorial: Mass customization manufacturing systems. *IEEE Transactions on Engineering Management*, 1(54), 4–11.

Boothroyd, G., Dewhurst, P., and Knight, W. A. (2010). *Product Design for Manufacture and Assembly*. Boca Raton: CRC Press.

Boysen, N., Fliedner, M., and Scholl, A. (2007). A classification of assembly line balancing problems. *European Journal of Operational Research*, 183(2), 674–693.

Brettel, M., Friederichsen, N., Keller, M., and Rosenberg, M. (2014). How virtualization, decentralization and network building change the manufacturing landscape: An Industry 4.0 perspective. *International Journal of Science, Engineering and Technology*, 8(1), 37, 44.

Cottyn, J., Van Landeghem, H., Stockman, K., and Derammelaere, S. (2011). A method to align a manufacturing execution system with lean objectives. *International Journal of Production Research*, 49(14), 4397–4413.

Da Silveira, G., Borenstein, D., and Fogliatto, F. S. (2001). Mass customization: Literature review and research directions. *International Journal of Production Economics*, 72(1), 1–13.

Davis, S. M. (1989). From "future perfect": Mass customizing. *Planning Review*, 17(2), 16–21.

Dombrowski, U., Wagner, T., and Riechel, C. (2013). Concept for a cyber physical assembly system. In *Assembly and Manufacturing (ISAM), 2013 IEEE International Symposium on* (pp. 293–296), Xi'an. IEEE.

Drunk, G. (2007). How much modularity needs the assembly? Ten years of modular platforms for assembly systems (in German: Wie viel Modularität braucht die Montage? Zehn Jahre modulare Plattformen für Montageanlagen), Report xpertgate.

Eastman, C. M. (2012). *Design for X: Concurrent Engineering Imperatives.* Dordrecht: The Netherlands. Springer Science & Business Media.

ElMaraghy, H. A., and Wiendahl, H. P. (2009). Changeability: An introduction. In ElMaraghy, H. A. (Ed.), *Changeable and Reconfigurable Manufacturing Systems* (pp. 3–24). London: Springer.

Fisher, M. L., and Ittner, C. D. (1999). The impact of product variety on automobile assembly operations: Empirical evidence and simulation analysis. *Management Science*, 45(6), 771–786.

Hart, C. W. (1995). Mass customization: Conceptual underpinnings, opportunities and limits. *International Journal of Service Industry Management*, 6(2), 36–45.

Hernandez, G., Allen, J. K., and Mistree, F. (2006). A theory and method for combining multiple approaches for product customization. *International Journal of Mass Customization*, 1(2–3), 315–339.

Hu, S. J., Ko, J., Weyand, L., El Maraghy, H. A., Lien, T. K., Koren, Y., Bley, H., Chryssolouris, G., Nasr, N., and Shpitalni, M. (2011). Assembly system design and operations for product variety. *CIRP Annals: Manufacturing Technology*, 60(2), 715–733.

Huang, G. Q., Fang, M. J., Lu, H., Dai, Q. Y., Liu, W., and Newman, S. (2009). RFID-enabled real-time mass-customized production planning and scheduling. In *19th International Conference on Flexible Automation and Intelligent Manufacturing (FAIM)* (pp. 1091–1103), University of Bath.

Jakubowski, J., and Peterka, J. (2014). Design for manufacturability in virtual environment using knowledge engineering. *Management and Production Engineering Review*, 5(1), 3–10.

Kaplan, A. M., and Haenlein, M. (2006). Toward a parsimonious definition of traditional and electronic mass customization. *Journal of Product Innovation Management*, 23(2), 168–182.

Keddis, N., Kainz, G., Buckl, C., and Knoll, A. (2013). Towards adaptable manufacturing systems. In *Industrial Technology (ICIT), 2013 IEEE International Conference on* (pp. 1410–1415), Cape Town: IEEE.

Koren, Y. (2005). Reconfigurable manufacturing and beyond (keynote paper). In *Proceedings of CIRP 3rd International Conference on Reconfigurable Manufacturing.* Ann Arbor, MI.

Koren, Y., Hu, S. J., Gu, P., and Shpitalni, M. (2013). Open-architecture products. *CIRP Annals: Manufacturing Technology*, 62(2), 719–729.

Koste, L. L., and Malhotra, M. K. (1999). A theoretical framework for analyzing the dimensions of manufacturing flexibility. *Journal of Operations Management*, 18(1), 75–93.

Kramer, J., Noronha, S., and Vergo, J. (2000). A user-centered design approach to personalization. *Communications of the ACM*, 43(8), 44–48.

Kull, H. (2015). Intelligent manufacturing technologies. In *Mass Customization: Opportunities, Methods, and Challenges for Manufacturers* (pp. 9–20). New York: Apress.

Lee, H. L., and Billington, C. (1994). Designing products and processes for postponement. In Dasu, S. and Eastman, C. (Eds.), *Management of Design* (pp. 105–122). New York: Springer Science and Business Media.

Lefever, D., and Wood, K. (1996, August). Design for assembly techniques in reverse engineering and redesign. In *ASME Design Theory and Methodology Conference* (pp. 78712–1063). ASME IDETC (International Design Engineering Technical Conference).

Liu, M., Ma, J., Lin, L., Ge, M., Wang, Q., and Liu, C. (2014). Intelligent assembly system for mechanical products and key technology based on Internet of Things. *Journal of Intelligent Manufacturing*, 1–29.

MacDuffie, J. P., Sethuraman, K., and Fisher, M. L. (1996). Product variety and manufacturing performance: Evidence from the international automotive assembly plant study. *Management Science*, 42(3), 350–369.

Matt, D. T., Rauch, E., and Dallasega, P. (2015). Trends towards distributed manufacturing systems and modern forms for their design. *Procedia CIRP*, 33, 185–190.

Matt, D. T., Rauch, E., and Franzellin, V. (2014). SMART reconfigurability approach in manufacture of steel and façade constructions. In Zaeh, M. F. (Ed.), *Enabling Manufacturing Competitiveness and Economic Sustainability* (pp. 29–34). Springer.

Modrak, V., Marton, D., and Bednar, S. (2015). The influence of mass customization strategy on configuration complexity of assembly systems. *Procedia CIRP*, 33, 539–544.

Mourtzis, D., Doukas, M., and Psarommatis, F. (2013). Design and operation of manufacturing networks for mass customisation. *CIRP Annals: Manufacturing Technology*, 62(1), 467–470.

Nambiar, A. N. (2009, July). Mass customization: Where do we go from here? In Ao, S. I., Gelman, L., Hukins, D. W. L., Hunter, A., and Korsunsky, A. M. (Eds.), *Proceedings of the World Congress on Engineering* (Vol. 1, pp. 687–693). Hong Kong: Newswood

Pine, B. J. (1993). *Mass Customization: The New Frontier in Business Competition.* Boston: Harvard Business Press.

Qiao, G., Lu, R. F., and McLean, C. (2006). Flexible manufacturing systems for mass customisation manufacturing. *International Journal of Mass Customisation*, 1(2–3), 374–393.

Reichwald, R., and Piller, F. (2005). Open innovation: Kunden als Partner im Innovationsprozess (Customers as partners in the innovation process). Electronically published: https://www.researchgate.net/profile/Frank_Piller/publication/235700667_Open_Innovation_Kunden_als_Partner_im_Innovationsprozess/links/0deec52c78a7d3b478000000.pdf, accessed on 9.8.2016.

Salonitis, K. (2014). Modular design for increasing assembly automation. *CIRP Annals: Manufacturing Technology*, 63(1), 189–192.

Schmitt, R., Brecher, C., Corves, B., et al. (2012). Self-optimising production systems. In Brecher, C. (Ed.), *Integrative Production Technology for High-Wage Countries* (pp. 697–986). Berlin: Springer.

Schreier, M. (2006). The value increment of mass-customized products: An empirical assessment. *Journal of Consumer Behaviour*, 5(4), 317–327.

Smirnov, Y. (1999). Manufacturing planning under uncertainty and incomplete information. In *American Association for Artificial Intelligence Spring Symposium*, Stanford (pp. 125-129). AAAI Technical Report SS-99-07.

Stoetzel, M. (2012). Exploiting mass customization towards open innovation. In *Proceedings 5th International Conference on Mass Customization and Personalization in Central Europe*. Novi Sad, Serbia: MCP-CE.

Terkaj, W., Tolio, T., and Valente, A. (2009). A review on manufacturing flexibility. In Tolio, T. (Ed.), *Design of Flexible Production Systems* (pp. 41–61). Berlin: Springer.

Thirumalai, S., and Sinha, K. K. (2011). Customization of the online purchase process in electronic retailing and customer satisfaction: An online field study. *Journal of Operations Management*, 29(5), 477–487.

Tseng, M. M., and Hu, S. J. (2014). Mass customization. In Laperrière, L., and Reinhart, G. (Eds.), *CIRP Encyclopedia of Production Engineering* (pp. 836–843). Berlin: Springer.

Whitney, D. E. (1988). Manufacture by design. *Harvard Business Review*, 66, 83–91.

Wiendahl, H.-P., El Maraghy, H. A., Nyhuis, P., Zäh, M. F., Wiendahl, H. H., Duffie, N., and Brieke, M. (2007). Changeable manufacturing-classification, design and operation. *CIRP Annals: Manufacturing Technology*, 56(2), 783–809.

Wiendahl, H.-P., Reichardt, J., and Nyhuis, P. (2009). *Handbuch Fabrikplanung: Konzept, Gestaltung und Umsetzung wandlungsfähiger Produktionsstätten* (Manual factory planning: Concept, design and implementation of agile production). Munich: Carl Hanser.

Xiaosong Peng, D., Liu, G., and Heim, G. R. (2011). Impacts of information technology on mass customization capability of manufacturing plants. *International Journal of Operations and Production Management*, 31(10), 1022–1047.

Zaeh, M. F., Beetz, M., Shea, K., Reinhart, G., Bender, K., Lau, C., Ostgathe, M., et al. (2009). The cognitive factory. In ElMaraghy, H. (Ed.), *Changeable and Reconfigurable Manufacturing Systems* (pp. 355–371). London: Springer.

Zhang, Y. F., Zhang, G., Wang, J. Q., Sun, S. D., Si, S. B., and Yang, T. (2014). Real-time information capturing and integration framework of the Internet of manufacturing things. *International Journal of Computer Integrated Manufacturing*, 28(8), 811–822.

Zhong, R. Y., Dai, Q. Y., Qu, T., Hu, G. J., and Huang, G. Q. (2013). RFID-enabled real-time manufacturing execution system for mass-customization production. *Robotics and Computer-Integrated Manufacturing*, 29(2), 283–292.

Zhu, X., Hu, S. J., and Marin, S. P. (2008). Modeling of manufacturing complexity in mixed-model assembly lines. *Journal of Manufacturing Science and Engineering*, 130(5), 051013–051013-10. doi:10.1115/1.2953076.

chapter three

Role of information systems in mass customization

Arun N. Nambiar

Contents

ABSTRACT

The dawn of the globalization era has been both a boon and a bane for companies and enterprises engaged in providing goods and services. Globalization opened up new markets, thus allowing companies to expand beyond their traditional market base. It also provided access to new sources of raw materials, the latest technologies, and a skilled workforce. However, it also engendered stiff competition, with too many players vying for the same market base. Seeing the shift, customers began to demand products and services customized to their needs, thus making customers truly the kings and queens. As a result, with the help of practitioners and researchers, companies continuously strive to find new ways to provide the product/service mix that will attract customers while keeping the costs down. Mass customization is a paradigm that focuses on providing customers with a customized experience at a reasonably low cost. This paradigm requires significant data collection across the entire operation of the enterprise and extensive analysis of the collected data to identify information related to customer preferences and how they translate into product or service features. With the advances in information technology in the recent years (also a result of globalization), it is possible to leverage information systems to facilitate this data collection and information processing. This chapter will identify the key elements and features of a holistic mass customization information system that encompasses all activities involved in the customization process.

3.1 Introduction

Companies the world over are continuously striving to generate profits through increased market share and reduced operating costs. With the help of advances in technology and access to global markets and workforces, more and more companies are entering the milieu, resulting in increased competition. Gone are the days when customers were happy with mundane and uniform off-the-shelf products. Customers are becoming ever more demanding, and the stiff competition as a result of globalization creates a perfect market condition for the customer to be truly king or queen. This serves as an impetus for both practitioners and researchers to come up with ways to reduce costs and provide better value to the customer. It has been shown (Agouridas et al., 2001) that the value of a product as perceived by the customer has evolved over the years from being focused on price and quality to including customized service, ease of placing an order, and quick order turnaround among other aspects of the product. Thus, there is increased emphasis on the value of the product and a renewed focus on the concept of the *value chain* (Porter, 1985). Focusing on the value chain forces companies to examine each activity that is carried out within its enterprise in the light of its value to the product or service being offered.

With customers considering the ability to customize products or services a significant part of their value, enterprises quickly adapted to the new scenario by introducing a plethora of choices for customers to choose from in the hope that some or all of their offerings might attract customers. However, the profusion of choices only served to confound the customer further, since neither did the customer understand the variety of features offered nor were they packaged appropriately for them to be of value to the customer. At the other end of the spectrum were companies that were providing individualized products, albeit at an exorbitantly high price, thus putting it out of reach for a large segment of the population.

Joseph Pine, often considered the father of mass customization, defined mass customization as "providing tremendous variety and individual customization, at prices comparable to standard goods and services ... with enough variety and customization that nearly everyone gets exactly what they want" (Pine et al., 1993; Pine, 1999). As the definition suggests, the idea is to provide customers their version of the product. This is a tectonic shift from the typical one-size-fits-all approach that companies have been practicing so far. The impetus for this change is competition and the need to gain market share in order to stay profitable. It has been shown (Jiao et al., 2001) that this approach provides better value to customers. However, it has also been shown (Svensson and Barford, 2002) that high operating cost is one huge deterrent.

Computers have been successfully leveraged by many manufacturing companies since the 1950s (Cooper, 1957). Some of the initial uses of computers were in the areas of scheduling and production planning (Kocchar, 1978, 1981; Kimber, 1988). Computers have been used mainly for brute force analysis or to solve mathematical models to determine optimal solutions to manufacturing and scheduling problems. Simulation (Yang and Kou, 2009) has also helped organizations evaluate alternative solutions for process improvements. As organizations grew and enterprises became more geographically dispersed, there arose a need for system-level information systems. Solutions such as *materials resource planning* (MRP) and *enterprise resource planning* (ERP) software became more popular, allowing companies to leverage information technology to coordinate its dispersed activities. In recent years, cloud computing has become widely popular, where idle computing resources from around the world are harnessed to solve computationally intensive problems. No longer need companies be tied down by legacy computing equipment or a lack of resources to make huge capital investments in expensive hardware. Numerous solutions are offered now as a service over the cloud, thus allowing companies to leverage the network to achieve its processing and computing needs. Extending the concept of cloud computing over to manufacturing, companies are beginning to seek out manufacturing resources on a need basis (Husejnagic and Sluga, 2015) instead of investing in high-end capital equipment.

Information systems facilitate data capture, storage, analysis, and retrieval. All of these features are critical in a mass customization environment to manage customer preferences and their relation to product features in a cost-effective manner. In fact, information systems are one of the two pillars of mass customization (Pine, 1999). Some of the typical features of an information system in mass customization include knowledge management and design configuration. This chapter will explore the role of information systems in the efficient and successful implementation of mass customization. Additionally, this chapter will identify some of the characteristics of a typical effective and efficient information system.

3.2 Background

Mass customization is often compared with mass production, since the prevailing practice is to produce goods and services on a large scale to benefit from the economies of scale. The mass production system was pioneered by Henry Ford (Ford and Crowther, 1922) in the 1920s, when large quantities of limited products were manufactured and assembled. Customers had limited, if any, choice in the products and services. A mass production system is typically vertical in nature, where each partner aims to mass-produce goods for the customer downstream. There is very limited information sharing between the partners in the value chain beyond the immediate supplier–customer pair. This results in information failure or a lack of information, which can lead to significant problems. The well-known bullwhip effect (Lee et al., 1997), where demand information is distorted as it travels up the supply/value chain, is a cause of excess production and the resulting excess inventory and increased waste in many industries.

Mass customization, introduced by Stan Davis (1989) and developed by Joseph Pine (1999), requires a significantly different approach. The partners in a mass customization value chain need to be more cohesively integrated and need to work closely with each other in order to be able to deliver customized products at mass production prices. The key features of the mass customization paradigm include product differentiation, cost reduction, and responsiveness (Nambiar, 2009b; Ngniatedema, 2012). Thus, in stark contrast to mass production and other manufacturing paradigms (Nambiar, 2010), a mass customization system is considered to be horizontal, with increased collaboration and information sharing. Thus, the information system needs are vastly different (Dean et al., 2008; Ngniatedema, 2012) for both these systems.

Understanding the mind of a customer is critical for the success of mass customization. A typical customer goes through five stages while making a purchasing decision. These include (Dibb et al., 2001) identifying the need or problem, searching for more information about various products, evaluating all alternatives, making a purchase, and finally evaluating the product. This is particularly important for mass customization. If a company is able to interact with the customer at the first stage of identifying the need or problem, it can offer customized solutions to the need, thereby increasing the probability that the customer will evaluate the proposed alternative favorably and ultimately make the purchase.

It is also important to understand the factors that influence the customer in making the purchasing decision. These factors can be broadly classified into four categories (Turban and King, 2003). These include personal characteristics such as the age of the person and his or her level of education. Gender also plays a significant role in purchasing decisions. Environmental characteristics include society in general and what is considered acceptable in the society to which the customer belongs. Some customers might also be conscious of how they affect the environment and focus on sustainability aspects such as environmental pollution, recycling, and so on. Access to effective customer support is becoming more and more critical. This, together with the logistical aspects of how quickly the product will be delivered, constitutes the third category. The fourth category includes marketing and other external factors that influence the customer. Understanding these four categories of factors influencing the customer has a significant impact on the success of mass customization initiatives.

It can be seen that information about a customer's decision-making process and the factors that influence the decision-making process are crucial. All pertinent information needs to be collected and analyzed to identify patterns that will help the company develop customization options. Information efficiency has been identified as a critical success factor (Mahajan et al., 2002) for the retailing sector. With the evolution and ever-increasing popularity of online retailing, information becomes even more vitally important. It has been shown (Varadarajan and Yadav, 2002) that the online environment combines the benefits of traditional retailing through its information richness while lowering the mismatching of information between various players in the supply chain. There is a related concept called *collaborative commerce* or *c-commerce* (Lim, 2003), where partners in the value chain leverage the Internet to improve information exchange.

A system that integrates information from a wide variety of sources, such as customers and supply chain partners, including retailers, is indispensable (Reichwald et al., 2000) for a successful mass customization campaign. This system should be able to integrate the different modes of operation as identified by MacCarthy et al. (2003), such as order processing, design, production, and postprocessing. Peng et al. (2011) identify four main areas where information technology can help with mass customization implementation. These include product configuration, product development, manufacturing, and supplier coordination. The needs of such an integrated system have been nicely summarized by Frutos and Borenstein (2004) as follows: "Provide direct links among main agents involved in the customization process, namely customer, company and supplier." The needs of such as an information system vary based on the type of industry and type of process used. Similar integrated systems have been proposed for doors and windows (Dean et al., 2008), online retailing (Vrechopoulos, 2004), the housing industry (Shin et al., 2008), the shoe industry (Dietrich et al., 2007), and software development (Karpowitz et al., 2008).

3.3 Mass customization information systems

An information system architecture can be defined as (Cook, 1996) "a conceptual framework that includes the identification of different components of the information systems environment and their interrelationships." However, the needs of the information systems are different in a mass production environment and a mass customization environment. It has been suggested (Ngniatedema, 2012) that the information systems architecture for a mass customization environment should adopt a more horizontal structure based on the processes within the organization, thus facilitating greater integration. There have been numerous piecemeal approaches (Fulkerson, 1997; Da Silveira et al., 2001; Piller, 2002; Frutos and Borenstein, 2004; Zahed and Reddy, 2004; Yao et al., 2007; Chen et al., 2008; Dean et al., 2009) to meet specific information system needs in a mass customization environment. This is primarily because of the diverse nature of the industry, making the development of a single one-size-fits-all architecture an arduous task. Another interesting approach (Verdouw et al., 2010) has been to apply the concept of mass customization itself to developing such systems by plugging together individual modules as may be needed. The goals of such a holistic information system architecture for mass customization (Ngniatedema, 2012) include

- Integrating data across the entire system
- Presenting information to the customer in a clear and lucid manner
- Analyzing copious amounts of data to garner knowledge about customer preferences
- Keeping up with the technological advances
- Reducing information distortion

Barrenechea (2010) developed a suitability index to determine if a given information system is suitable to the company's needs based on a series of criteria. This index allows companies to gage their level of preparedness vis-à-vis information systems for mass customization initiatives. Despite the subtle variations across companies and industries regarding information system needs in a mass customization environment, certain key elements of the architecture are indispensable irrespective of the industry or size of the company. These will now be described.

3.3.1 Customer relationship management

The success of mass customization hinges on the ability of companies to capture and maintain customer preferences. Strauss and Frost (2001) identify the three steps involved in a typical customer relationship management (CRM) process. This includes identification, differentiation, and customization. It can been seen that all three steps are equally important for the mass customization initiative as well. It is important to identify the customer needs, differentiate customers into different groups based on their preferences, and customize the offerings for each group. Even though CRM has been around for a long time, its application and use in today's world of Internet and mobile technology is not as widespread (Feinberg et al., 2002). This allows potential developers to integrate CRM with mass customization systems, since both systems are so interdependent.

3.3.2 Product configuration systems

The very essence of mass customization is the ability to provide highly customized products or services. However, as the number of product or service features increase, the number of possible combinations increase exponentially, thus making it virtually impossible for a company to provide all of its customers with off-the-shelf products or services tailored to their needs. Instead, companies provide customers with the ability to design their own product or service. This provides a truly customized experience for customers, who benefit from receiving a product or service that meets their specific needs. However, this becomes an onerous task given the plethora of choices that companies are wont to give their customers. Thus, a configuration system (Forsza and Salvador, 2006) that allows customers and product designers to design a product or service of their choice based on the available features is an integral part of the information system needs of a mass customization company. The system needs to not only provide customers with the ability to design their own products or services but also provide preconfigured systems based on inferred customer choices, thus making the customer's task easier. The configuration system also needs to provide customers with the relevant information so that they can make informed decisions regarding their product or service choices. This information needs to be provided in a clear and lucid format bereft of technical jargon so that it caters to the disparate backgrounds of its customers.

The scope of such a product configurator (Forza and Salvador, 2006) varies based on the type of business practice adopted by the company. At one end of the spectrum is the practice where a company provides variety with little customization, where the customer does not interact with the configurator at all. For example, consider the case of Tropicana orange juice (Tropicana, 2016). There is a wide range of orange juices available from the same company, such as juice with pulp, without pulp, fortified with vitamins or fiber, and so on. In this case, the customer does not design the product directly. Instead, the company designs the product based on customer input through surveys and questionnaires. At the other end of the spectrum is a truly customized approach, where the customer is involved right from the design stage. For example, consider the case of custom-made Rolls-Royce cars (Rolls-Royce, 2016), where the product configurator allows the customers to build their own cars. The success of a product configurator also hinges on efficiently translating customer needs into features in the product or service. The more efficient this translation, the smoother the customer experience.

3.3.3 Product design

Product or service design needs to be based on modularity (Magrab, 1997), thus allowing customization based on customer preferences. Consider the example of laptop computers. Most companies, such as Dell, Apple, or Lenovo, allow customers to custom-build their laptop on their website. Each of the myriad features of a laptop—such as the hard drive, memory, or screen size, to name a few—are individual modules that can be plugged together to make the customized computer. The need for a good product data management system has been underscored by many researchers (Agouridas et al., 2001; Pan et al., 2014). As the number of features and options increase, the total number of parts and assemblies that need to be tracked increase exponentially (Jiao and Tseng, 1999), thus making inventory management an onerous task if carried out individually by the partners. This necessitates a fully networked system that integrates inventory from all partners in the value chain into a holistic system that allows for more efficient inventory management.

3.3.4 System integration

Information systems have been around for a long time, and the underlying technology has been evolving at a drastic pace over the years. As new technologies and tools are developed, new systems are designed as well. Moreover, due to the open-source nature of some of the technologies, it becomes easy for entrepreneurs to set up companies that are engaged in developing software systems using these technologies. This proliferation of software offerings and the underlying technologies provides companies with a wide variety of choice to accomplish their information system needs. As a company grows, its information system needs change as well. Moreover, in today's age, a company often does not exist by itself; it has numerous partners along its supply chain. This compounds the complexity of the information system. Thus, there are three main requirements for the successful company-wide integration of an information system.

- *Legacy systems integration*: Due to the very nature of technology and the fast pace at which changes occur in this field, companies are often forced to upgrade to newer systems in order to keep up with competition and customer needs. As a result, companies are saddled with multiple versions of software systems that store data in diverse formats. As data is key in mass customization, it is vital that newer systems are able to interact with these legacy systems and leverage historical data to provide better customer choice. Efficient data exchange mechanisms are integral to this seamless interaction with legacy systems. These mechanisms make data portable across diverse systems, thus enabling companies to make effective use of all the resident data.

- *Cross-technology integration*: It is essential that all partners in the supply chain are able to communicate and seamlessly exchange information with each other in order for the company to be truly agile and responsive to changing customer needs. Due to the varying needs of the supply chain partners, each partner may choose to have its own software system to support its information system needs. With the plethora of systems available and the diverse nature of underlying technology, software systems in these supply chain partners have to be able to communicate across platforms and technologies. This calls for an effective mechanism for porting data into different formats that are readable by these disparate systems.
- *Seamless integration*: Oftentimes, the end users of these software systems are not technologists conversant with the inner workings of these systems. Also, given the multitude of platforms and technologies, it would be impractical to expect the end user to be able to manipulate the innards of the software system to allow for communication with legacy and diverse software systems. Thus, along with data portability and legacy integration, the company's information system needs to be able to achieve this integration seamlessly so that the end user can focus on data analysis and design development.

3.3.5 Data capture mechanisms

Since information is one of the two pillars of mass customization and data is the foundational basis for information, collecting and storing data is a significant activity in ensuring the success of mass customization efforts. Data can be collected through various direct and indirect mechanisms. In *direct mechanisms*, the product, service, or customer is actively involved in the data collection process. Some direct mechanisms include

- *Point of sale (POS)*: As products are being checked out at the store (or purchased in an online environment), information about the sale is recorded into the system. This information can help monitor stock and trigger production and/or deliveries based on the inventory levels.

- *Identification tags*: Radio frequency identification (RFID) tags are also becoming increasingly popular to help with inventory management and data collection. These tags have a unique advantage over bar codes in that they do not require a direct line of sight with the scanner, thus drastically speeding up the process of stocktaking.
- *Customer surveys*: Customizing products or services relies heavily on customer input. Customer input can be obtained in numerous ways, such as survey questionnaires, interviews, sampling or trial offers, and so on. The feedback obtained from these instruments can be utilized to ascertain customer preferences and thus steer product or service design in the right direction.

Indirect mechanisms, as the term suggests, do not require the active involvement of the product, service, or customer. Customer behavior is closely monitored to ascertain their preferences. These mechanisms have become more relevant these days with the proliferation of online shopping. Some of the indirect mechanisms for collecting data include

- *Online visitors*: The number of people browsing the company's product page or news articles can also be mined to gage the types of customers and the level of customer interest in a new product/service or a new variation of an existing product/service. It has been shown that (Turban and King, 2003) understanding the customers and providing a tailored experience is critical for the success of companies in an online environment. For example, if a company has released news about its upcoming new product/service, the number of unique visitors or readers might help gage the level of interest among the customers.
- *Customer profiles*: Many websites require customers to open an account and create a profile before utilizing the services. Every product or service that the customer orders is automatically collected and stored in the customer's profile. Through an in-depth analysis of this data, companies can identify customer preferences and tailor their products or services based on these preferences. Understanding customer behavior (Newman et al., 2002) is critical for the success of the customization efforts. For example, if a customer consistently books an aisle seat on an airline website, the airline company could automatically tailor the choices to offer more aisle seats, thus customizing the experience based on the preferences.

Before diving into installing mechanisms to collect data, however, it is important to analyze what data is required and what are the best mechanisms to collect and store this data (O'Brien, 2002). It is also important not to confuse customization with personalization (Strauss and Frost,

2001). Customization is the ability to provide a mix of products based on user preferences, while personalization is adding a personal touch to the interaction by including the customer's name and so on. Moreover, some instruments might be more appropriate than others, depending on the type of interaction within the supply chain, such as the supplier–retailer relationship, the retailer–consumer relationship, or in some rare cases, even the supplier–consumer relationship (Vrechopoulos, 2004). Using these instruments to tackle the three previously mentioned steps in the CRM process can work wonders in enhancing the success of the mass customization approach.

3.3.6 Supplier relationship management

A big part of today's value chain in any industry is the array of suppliers and vendors that provide a wide variety of services, such as raw materials, processing, packaging, or even distribution. Companies are offloading routine activities to suppliers in order to focus on the niche areas for which they are known. For example, Apple is known for its immaculately designed and exquisitely engineered products. Thus, the company focuses on design and engineering while engaging subcontractors for the actual assembly. Some companies took this approach overboard and pitted suppliers and contractors against each other to lower costs, thus resulting in an ugly bidding battle, leading to corners being cut to save on cost, which in turn engendered quality issues. Toyota, which pioneered the Toyota Production System, focusing on the elimination of waste, ran its operations with a handful of suppliers with whom it had long-standing relationships. This long-term relationship begat trust, which in turn allowed the suppliers and Toyota to work together and share information related to product, design, and production, automatically leading to lowered costs and improved quality. It has been shown (Liao et al., 2011) that trust is essential for greater information sharing and increased collaboration, which is indispensable for a mass customization environment.

3.3.7 Data analytics

Information is key to the success of mass customization, and this information is obtained by sifting through data in the form of sales orders, purchase history, customer preferences, and needs. For example, consider the airline industry. By analyzing the seating choices made by a frequent flyer, the information system used to book tickets should be able to determine the most likely seating preference for the customer, thus providing the customer with customized seating. The information system should be able to collect data and leverage the collected data to determine desirable product or service features. However, this is not a simple task, given that in a truly connected system, there are copious amounts of data being collected every day. Analyzing this data manually is an onerous and practically impossible task. Thus, the information system needs to have the capability to analyze data to identify patterns in customer preferences.

3.3.8 User experience

With the widespread availability of the Internet, and more and more business being conducted on mobile devices, it is imperative that businesses adapt to this changing scenario and provide stakeholders and customers with the ability to seamlessly interact with the information system irrespective of the device used. This again represents a dramatic shift from the traditional desktop-based systems. There are two main approaches to providing this seamless interaction.

- *Responsive design*: In this approach, the main portal through which customers interact with the system reconfigures itself based on the device being used and the resulting screen resolution. This is achieved through behind-the-scenes web programming that understands the device and reorganizes the elements on the web page based on the screen resolution of the device. This approach has the advantage of providing customers with a uniform experience across devices. It also makes the task of system maintenance and upkeep easier since there is only one system to manage. However, this approach does not fully leverage the capabilities of the mobile devices, and companies often develop a separate stand-alone application that its customers can use to interact with through their mobile devices.
- *Mobile app*: In this approach, a stand-alone application is downloaded to the customer's device, which then provides the necessary interaction. This allows companies to leverage the capabilities of a truly mobile device. This, however, compounds the task of maintaining these systems since there needs to be a separate application for each platform, such as Apple's iOS, Google's Android, and so on. However, despite this hurdle, many companies choose this approach for multiple reasons.
  - *Customer perception*: Customers might perceive the company to be on the forefront of technology if it has a mobile app. This helps with building the company's reputation and allows the company to leverage this into improved market share.
  - *Customer experience*: The mobile applications developed for each platform can make use of the unique features available in these platforms to provide customers with a better experience.

Apart from these key elements, the mass customization information system also needs to incorporate the following features:

- *Adaptability and flexibility*: Markets are very dynamic in nature and customer preferences tend to be very fluid, constantly changing over time. Thus, it is imperative that the information system be agile (Nambiar, 2009a) and have the capability to be reconfigured to respond to changes in the market dynamics (*flexibility*). It is also important to be proactive and envision some of the changes in the market or to even create watershed moments that completely revolutionize the market (e.g., Apple's iPhone). A truly *adaptable* system will be able to seamlessly handle these changes with minimal impact on the existing operations. It has been shown (Porter et al., 1999) that a lack of reconfigurability is often a significant factor for the reluctance on the part of companies to invest in such systems.
- *Scalability*: A successful company grows over time and diversifies its product mix to capitalize on its reputation and enter into new markets. Any chosen information system should be able to grow with the company and accommodate its information-processing needs. However, there will potentially come a time when the information system is bursting at its seams and unable to handle the ballooning needs of the enterprise. This is where interoperability becomes crucial.
- *Interoperability*: As companies come together with the common objective of providing a truly customized experience for the customer, as is wont these days, they may already have a functioning information system that handles some if not all aspects of their operations. It would be redundant to require all partners to reinvest in another completely new system. Thus, any implementation of information systems needs to be interoperable with other legacy and existing systems. This facilitates free and seamless information sharing, which is an essential facet of mass customization.

3.4 Future research directions

Analysis of the literature has revealed a renewed interest among practitioners and researchers in this area of leveraging information systems for the successful implementation of the mass customization paradigm. Although extensive work has been done in certain areas of the customization value chain, such as developing smart and efficient product configurators and managing customer relationships, to name a few, there is still work to be done in developing a holistic information management system that fully integrates all facets of the customization process. Standards need to be developed for coding information so that it can be easily shared across partners (Dean et al., 2008). Integrating legacy systems and coalescing piecemeal technologies into a comprehensive system also need to be investigated further. Knowledge management (Nambiar, 2013) is another area that is gaining in importance in order to manage the resident tacit knowledge about customer preferences and product features. Integrating this into the mass customization framework (Helms et al., 2008) can help leverage the knowledge to provide a better customized experience.

3.5 Conclusions

The mass customization paradigm holds significant promise for companies stifled by competition and a lack of new products or services. Despite being around for more than two decades, this concept is still lagging behind in its practical implementations. This is primarily due to the complexities involved in successfully putting into practice the tenets of mass customization. Information technology can be leveraged extensively to help overcome some of this implementation hurdle. This has already been underscored by the numerous piecemeal approaches developed to address one or more aspects of the process. However, a fully integrated information system with the key elements and features identified in this chapter can go a long way to improving the success of mass customization implementation, thus allowing this paradigm to really take wing and become more commonplace, like its lean counterpart.

References

Agouridas, V., Allen, M., McKay, A., Pennington, A., and Holland, S. (2001). Using information structures to gain competitive advantage. *Proceedings of the 2001 PICMET: Portland International Conference on Management of Engineering and Technology*, Portland, OR, Vol. 2, pp. 681–692. IEEE.org.

Barrenechea, O. L. (2010). An information technology suitability index for mass customization. Thesis, University of Texas—Pan American. Ann Arbor, MI: ProQuest Dissertations Publishing.

Chen, R. S., Tsai, Y. S., and Tu, A. (2008). An RFID-based manufacturing control framework for loosely coupled distributed manufacturing system supporting mass customization. *IEICE Transactions on Information and Systems*, E91D(12), 2834–2845.

Cook, M. (1996). *Building Enterprise Information Architectures: Reengineering Information Systems*. Upper Saddle River, NJ: Prentice-Hall.

Cooper, A. H. (1957). Integration of computers with factory processes. *Journal of the British Institution of Radio Engineers*, 17(8), 431–441.

Da Silveira, G., Borenstein, D., and Fogliatto, F. (2001). Mass customization: Literature review and research directions. *International Journal of Production Economics*, 72(1), 1–13.

Davis, S. (1989). From "future perfect": Mass customizing. *Planning Review*, 17(2), 16–21.

Dean, P. R., Tu, Y. L., and Xue, D. (2008). A framework for generating product production information for mass customization. *International Journal of Advanced Manufacturing Technology*, 38(11–12), 1244–1259.

Dean, P. R., Tu, Y. L., and Xue, D. (2009). An information system for one-of-a-kind production. *International Journal of Production Research*, 47(4), 1071–1087.

Dibb, S., Simkin, L., Pride, W. M., and Ferrell, O. C. (2001). *Marketing: Concepts and strategies*. Fourth European Edition. Boston, MA: Houghton Mifflin.

Dietrich, A. J., Kirn, S., and Sugumaran, V. (2007). A service-oriented architecture for mass customization: A shoe industry case study. *IEEE Transactions on Engineering Management*, 54(1), 190–204.

Feinberg, R. A., Kadam, R., Hokama, L., and Kim, I. (2002). The state of electronic customer relationship management in retailing. *International Journal of Retail and Distribution Management*, 30(10), 470–481.

Ford, H., and Crowther, S. (1922). *My Life and Work*. Garden City, NY: Doubleday, Page.

Forsza, C., and Salvador, F. (2006). *Product Information Management for Mass Customization: Connecting Customer, Front-Office and Back-Office for Fast and Efficient Customization*. New York: Palgrave Macmillan.

Frutos, J. D., and Borenstein, D. (2004). A framework to support customer–company interaction in mass customization environments. *Computers in Industry*, 54, 115–135.

Fulkerson, B. (1997). A response to dynamic change in the market place. *Decision Support Systems*, 21(3), 199–214.

Helms, M. M., Ahmadi, M., Jih, W. J. K., and Ettkin, L. P. (2008). Technologies in support of mass customization strategy: Exploring the linkages between eCommerce and knowledge management. *Computers in Industry*, 59, 351–363.

Husejnagic, D., and Sluga, A. (2015). A conceptual framework for a ubiquitous autonomous work system in the engineer-to-order environment. *International Journal of Advanced Manufacturing Technology*, 78, 1971–1988.

Jiao, J., Ma, Q., and Tseng, M. M. (2001). Towards high value-added products and services: Mass customization and beyond. *Technovation*, 23, 149–152.

Jiao, J., and Tseng, M. M. (1999). A methodology of developing product family architecture for mass customization. *Journal of Intelligent Manufacturing*, 10, 3–30.

Karpowitz, D. J., Cox, J. J., Humpherys, J. C., and Warnick, S. C. (2008). A dynamic workflow framework for mass customization using web service and autonomous agent techniques. *Journal of Intelligent Manufacturing*, 19(5), 537–552.

Kimber, A. (1988). Production planning and scheduling on process plant. *Process Engineering*, 69(2), 81–82.

Kocchar, A. K. (1978). Use of computers and analytical techniques for production planning and control in the British manufacturing industry. *Computers and Industrial Engineering*, 2(4), 163–179.

Kocchar, A. K. (1981). Low-cost computers in production planning and control systems. *Production Engineer*, 60(5), 44–47.

Lee, H. L., Padmanabhan, V., and Whang, S. (1997). Information distortion in a supply chain: The bullwhip effect. *Management Science*, 43(4), 546–558.

Liao, K., Ma, Z., Lee, J. J., and Ke, K. (2011). Achieving mass customization through trust-driven information sharing: A supplier's perspective. *Management Research Review*, 34(5), 541–552.

Lim, J. M. (2003). The impact of information technology on mass customization: An evolving trend for companies that want to be successful in the e-business arena. Thesis, Rochester Institute of Technology, NY.

MacCarthy, B., Brabazon, P. G., and Bramham, J. (2003). Fundamental modes of operation for mass customization. *International Journal of Production Economics*, 85, 289–304.

Magrab, B. E. (1997). *Integrated Product and Process Design and Development*. New York: CRC Press.

Mahajan, V., Srinivasan, R., and Wind, J. (2002). The dot.com retail failures of 2000: Were there any winners? *Journal of the Academy of Marketing Science*, 30(4), 474–486.

Nambiar, A. N. (2009a). Agile manufacturing: A taxonomic framework for research. *Proceedings of the 2009 International Conferences on Computers and Industrial Engineering*, 684–689. IEEE.org.

Nambiar, A. N. (2009b). Mass customization: Where do we go from here? *Proceedings of the 2009 World Congress on Engineering*, London, July 1–3, Vol. 1. International Association of Engineers.

Nambiar, A. N. (2010). Modern manufacturing paradigms: A comparison. *Proceedings of the 2010 International Multiconference of Engineers and Computer Scientists*, Hong Kong, March 17–19, Vol. 3. International Association of Engineers.

Nambiar, A. N. (2013). A current perspective of knowledge management in a global economy. *International Journal of Soft Computing and Software Engineering*, 3(3), 236–239.

Newman, A. J., Yu, D. K. C., and Oulton, D. P. (2002). New insights into retail space and format planning from customer-tracking data. *Journal of Retailing and Consumer Services*, 9(5), 253–258.

Ngniatedema, T. (2012). A mass customization information systems architecture framework. *Journal of Computer Information Systems*, 52(3), 60–70.

O'Brien, J. A. (2002). *Management Information Systems: Managing Information Technology in the e-Business Enterprise*, Fifth Edition. New York: McGraw Hill.

Pan, X., Zhu, X., Ji, Y., Yang, Yi., and Wu, Y. (2014). An information integration modelling architecture for product family life cycle in mass customisation. *International Journal of Computer Integrated Manufacturing*, 27(9), 869–886.

Peng, D. X., Liu, G., and Heim, G. R. (2011). Impacts of information technology on mass customization capability of manufacturing plants. *International Journal of Operations and Productions Management*, 31(10), 1022–1047.

Piller, F. T. (2002). Customer interaction and digitization: A structured approach to mass customization. In *Moving into Mass Customization: Information Systems and Management Principle—Part II*, pp. 119–137. Berlin: Springer.

Pine, B. J. (1999). *Mass Customization: The New Frontier in Business Competition*. Boston, MA: Harvard Business School Press.

Pine, B. J., Victor, B., and Bonyton, A. C. (1993). Making mass customization work. *Harvard Business Review*, 71(5), 108–119.

Porter, K., Little, D., Peck, M., and Rollins, R. (1999). Manufacturing classifications: Relationships with production control systems. *Integrated Manufacturing Systems*, 10(4), 189–199.

Porter, M. E. (1985). *Competitive Advantage*. New York: Free Press.

Reichwald, R., Piller, F. T., and Moeslein, K. (2000). Information as a critical success factor for mass customization. *Proceedings of the 2000 ASAC-IFSAM Conference*, Montreal.

Rolls-Royce (2016). Rolls-Royce homepage. Retrieved from https://www.rolls-roycemotorcars.com/en-GB/home.html. Accessed on January 2, 2016.

Shin, Y., An, S., Cho, H., Kim, G., and Kang, K. (2008). Application of information technology for mass customization in the housing construction industry in Korea. *Automation in Construction*, 17(7), 831–838.

Strauss, J., and Frost, R. (2001). *E-marketing*, Second Edition. Englewood Cliffs, NJ: Prentice-Hall.

Svensson, C., and Barford, A. (2002). Limits and opportunities in mass customization for "build-to-order" SMEs. *Computers in Industry*, 49, 77–89.

Tropicana (2016). Tropicana homepage. Retrieved from http://www.tropicana.com. Accessed on January 2, 2016.

Turban, E., and King, D. (2003). *Introduction to E-Commerce*. Englewood Cliffs, NJ: Prentice Hall.

Varadarajan, R. P., and Yadav, M. S. (2002). Marketing strategy and the Internet: An organizing framework. *Journal of Academy of Marketing Science*, 30(4), 296–312.

Verdouw, C. N., Beulens, A. J. M., Trienekens, J. H., and Verwaart, T. (2010). Towards dynamic reference information models: Readiness for ICT mass customization. *Computers in Industry*, 61, 833–844.

Vrechopoulos, A. P. (2004). Mass customization challenges in Internet retailing through information management. *International Journal of Information Management*, 24, 59–71.

Yang, J., and Kou, Y. (2009). Application of online simulation in production planning and scheduling. *Journal of Beijing University of Aeronautics and Astronautics*, 35(2), 215–218.

Yao, S., Han, X., Yang, Y., Rong, Y., Huang, S. H., Yen, D. W., and Zhang, G. (2007). Computer aided manufacturing planning for mass customization, part 3: Information modeling. *International Journal of Advanced Manufacturing Technology*, 32, 218–228.

Zahed, S., and Reddy, B. K. (2004). A mass customization information framework for integration of customer in the configuration design of a customized product. *Artificial Intelligence for Engineering Design, Analysis and Manufacturing*, 18, 71–85.

section two

Complexity drivers in mass customization

chapter four

Complexity issues in mass customized manufacturing

Hendrik Van Landeghem and El-Houssaine Aghezzaf

Contents

ABSTRACT

This chapter focuses on the role of complexity and its impact on the human operator within assembly-oriented mass customized manufacturing (MCM). It starts with a definition of the main complexity concepts we need for a thorough understanding of this chapter, building on the previous chapters. It also argues that automotive manufacturing is currently a good representative example of MCM, as the number of variants and models continue to proliferate, and it is one of the few mature industries that practice MCM. Their size and heritage forces them to look for and adopt structural and methodical approaches to deal with complexity.

Next, the major drivers of complexity and their impacts on manufacturing systems are described and analyzed. These were derived from operator debriefings on complexity and how they experience it during daily work. These findings are based on a 3-year international research project (2010–2012) involving leading original equipment manufacturers (OEMs) in automotive and vehicle manufacturing in Belgium and Sweden. One result that is described is a fast and objective measurement to determine the complexity of a manual assembly station, based on a statistical classification model.

The last part of the chapter is devoted to some of the methodical approaches that can be used to manage complexity in MCM; some of these approaches are based on recent research results from our research team at Ghent University. An enriched method to balance manufacturing lines, minimizing variability of station *takt* times, is discussed. The main objective is minimizing overloads that prove to be most detrimental to operator cognitive loads, quality issues, and assembly errors. Also, technological developments to enhance the information exchange with operators in a complex MCM setting are discussed, using wearable devices and bidirectional information exchange.

4.1 Complexity in assembly

The trend toward mass customization in the automotive industry and the complexity it induces has been the focus of many studies in the last three decades. Specifically, the impact of increased product variety on the performance of automotive mixed-model assembly lines has already been studied by Fisher et al. (1995), MacDuffie et al. (1996), and Fisher and Ittner (1999).

In particular, MacDuffie et al. (1996) investigated the effect of product variety on total labor productivity and consumer-perceived product quality in an international study covering 70 assembly plants, from 16 countries, participating in the International Motor Vehicle Program at MIT. In their paper, they analyzed complexity measures that capture different aspects of product mixes in assembly plants: *model mix complexity*, *parts complexity*, *option content*, and *option variability*. Model mix complexity is based on the number of different platforms, body styles, and models, scaled by the number of different body shops and assembly lines in each plant. Parts complexity is measured as an aggregate index of six factors:

- The number of variants for engines
- Idem for wire harnesses
- Idem for paint colors
- The total number of parts
- The percentage of common parts
- The number of suppliers

Option content reflects the overall level of installed options (from a list of 11), and equals the percentage of vehicles built with various options, aggregated across all models in a plant. Option variability captures the extent to which different cars have different amounts of options installed. A statistical analysis has shown significant but limited negative correlation between the complexity measures and the manufacturing performance. This seminal study has, however, already indicated the main factors that define complexity and the role of variability of those across the different products on the same line. The latter's importance has been reported in Fisher and Ittner (1999). Section 4.2 will explore this in more detail by constructing an influence model.

This work has been followed by many other investigations attempting to define, model, and develop valid and useful complexity measures for manufacturing systems. Frizelle (1996) suggested that a useful complexity measure needs to be separable and additive as it will then simplify its computation and allow easy analysis for managers. Deshmukh et al. (1998) thus provided a clear definition of *static* and *dynamic complexity*. Static or structural complexity is related to the structure of the system, the variety of components and products, the number of processes and machines, and so on. On the other hand, dynamic complexity measures the unpredictability in the behavior of the system over a time period.

Fujimoto and Ahmed (2001) proposed a complexity index based on the ease of assembling a product. The index takes the form of entropy to evaluate the ease of assembly of a product, defined as the uncertainty of gripping, positioning, and inserting parts in an assembly process. Workstation entropy is the sum of the variety coming from upstream and the variety added in the station itself. Fujimoto et al. (2003) also broadly identified some of the drivers and impacts of variety. Specifically, they identified the directions of approach as one of the drivers, which we have included in our complexity classification model in Section 4.3.

ElMaraghy and Urbanic (2003, 2004) proposed a methodology to assess product and process complexity and their interrelations in a systematic manner. A matrix methodology and an objective measure of complexity have been proposed that assess the three levels of manufacturing complexity: product complexity, process complexity, and operational complexity. Samy and ElMaraghy (2010) defined a product assembly complexity index. Their model considers not only the complexity of assembly introduced by factors from the design-for-assembly method but also the diversity of the various parts used in product assembly and the total part count. The proposed model incorporates the assembly complexity resulting from the number and diversity of the parts and fasteners used in the product assembly using a formulation that incorporates information content and diversity.

Zhu et al. (2008) and Hu et al. (2008), inspired by Fujimoto and Ahmed (2001), proposed a measure of manufacturing complexity at the workstation level introduced by product variety and modeled its propagation through the assembly system. The proposed complexity model quantifies human performance in making choices—that is, the uncertainty the operator is facing when making various choices at an assembly station.

- *Fixture choice*: Choose the right fixture according to the base part—that is, the partially completed assemblage to be mounted on as well as the added part to be assembled.
- *Tool choice*: Choose the right tool according to the added part to be assembled as well as the base part to be mounted on.
- *Procedure choice*: Choose the right procedure—for example, part orientation, approach angle, or the temporary unloading of certain parts due to geometric conflicts/subassembly stabilities.

Zhu (2009) also pointed out that complexity effects should be taken into account when determining the *assembly sequence*, as well as the *build sequence*. The assembly sequence is the (static) manner in which the assembly tasks are assigned to workstations along the line, while the build sequence is the (dynamic) sequence into which specific vehicle configurations (with their options and variants) are loaded onto the assembly line, typically in a daily schedule. The line-balancing method we describe in Section 4.4 addresses assembly sequencing.

In an attempt to understand complexity, Schuh et al. (2008) determined its main drivers as uncertainty, dynamics, multiplicity, variety, interactions, and interdependencies, and a combination of such proprieties that can render a system complex or not complex. Rodríguez-Toro et al. (2004) proposed a specific taxonomy where complexity is split into static and dynamic. Static complexity is associated with the product, whereas dynamic complexity is linked to the process.

Most of these published models seem to link complexity to the objective characteristics of products or processes. However, between complexity and the company's financial bottom line (measured through productivity and quality), we find the operator who has to perform the tasks at the workstation. It is intuitive that both complexity and time pressure play a factor in the mental workload of the operator. Parasuraman et al. (2000) provided a simplified but useful four-step interaction model of how humans deal with information from complex systems.

1. Information acquisition
2. Information analysis
3. Decision and action selection
4. Action implementation

Rasmussen (1983) provided an interesting model of how skilled operators deal with information and what the effect is of each mode of cognition on their performance. In it, he distinguishes three behavior levels: skill based, rule based, and knowledge based.

This entails the whole domain of ergonomics and the psychological side of task design. In Section 4.4, we will describe how operator information systems can mediate the effects of complexity on the cognitive load of the operator.

Very recently, ElMaraghy et al. (2012) published a very thorough literature review of complexity models in design and manufacturing. They state that "designing systems for less complexity [is an] important [issue] for further research." Our research has benefited greatly from insights gained from this research work and we will reference it throughout this chapter.

In Section 4.2, we present and discuss a first attempt to aggregate the various factors governing complexity in assembly and populate this generic structure with tangible elements. In Section 4.3, we describe objective measurement methods to determine the complexity of workstations. In Section 4.4, we present several approaches to cope with and reduce the effects of complexity in MCM situations.

4.2 Modeling the DNA of complexity

The concepts and models that are introduced and discussed in Sections 4.2 and 4.3 are the result of a 3-year international study that was conducted in the vehicle industry of Belgium and Sweden (De Lima Gabriel Zeltzer et al., 2013; Mattson et al., 2014). It included the major vehicle OEMs of these countries. The research focus on which this chapter is based relates to the workstations along driven assembly lines, where manual assembly work is carried out on different models in a mixed-model fashion. Different types of assembly lines were investigated, including two for car models and two for engine models, a truck model, and several subassembly lines with suppliers. This has led to a total of 76 different workstations from which data was gathered.

4.2.1 Complexity definition

A good definition of complexity has to be generic enough to be applicable to different manufacturing systems and at the same time specific enough to guide and support the decisions related to whether a system is complex or not. Although the literature review provided useful insights into manufacturing complexity, most approaches are relatively specific. In our view, there still existed a need for a clear, simple, and generic complexity definition. After an extensive exchange of ideas among the project partners, the following definition is proposed, which proved to be very useful during the workshops that led to the causal model: "The complexity of a workstation is the sum of all technical and ergonomical aspects and factors that make the set of tasks to be performed within it by an operator mentally difficult, error prone, requiring thinking and vigilance, and inducing stress." This definition recognizes the fact that the inherent complexity of tasks is determined to a large extent by the operator who executes them, hence the term *subjective complexity*. This means, according to the findings in this case, that the same set of tasks can be judged differently by different operators, production engineers, quality controllers, and line managers under different circumstances. This makes the issue of quantifying complexity in an unambiguous manner, so-called objective complexity, a real challenge (Mattson et al., 2014). One immediate consequence is that measuring the magnitude of subjective complexity will always involve a behavioral and psychological aspect, which is difficult to quantify. Secondly, since complexity is a multifaceted concept, it is almost impossible to measure it directly as no meaningful scale exists. In order to gain more insight into the nature of complexity, we set out to construct a causal model of complexity.

4.2.2 A causal model of complexity in manufacturing

ElMaraghy et al. (2012) proposed a causal map of how manufacturing complexity cascades down from product design to the cognitive and physical effort of the individual operator. This scheme inspired us to conduct a series of fact-finding workshops with vehicle manufacturing companies to validate and refine this causal map.

To gather as much useful information as possible, it was decided that the participants in these workshops should include shop floor employees, production engineers, quality controllers, and line managers—that is, all those who deal with complexity in their daily activities. In the first phase, the project objectives were explained to all participants. Next, the participants were asked to identify two low- and two high-complexity workstations. The participants were then asked to use these workstations as a mental reference while providing elements of complexity, drivers or causes of complexity, and impacts or consequences of complexity. Each participant provided his or her views individually by noting them on separate sticky notes. Afterward, the answers were processed in three rounds. The relationships between these rounds were argued and modeled into a causal model.

The results of the first round focused on aspects characterizing complexity. The results of the second round (*impacts*) concentrated on revealing the consequences of complexity: areas that are affected by complexity and the influence of complexity on the manufacturing activity and on the teams. The third round aimed at detecting the *direct drivers* of complexity—that is, the variables that are directly linked to the complexity elements as causal factors. In each round, the relevant notes from the participants were put up on a wall, clustered by similarity. Finally, a brainstorming session was held where the list of ideas was discussed and finalized. The results were discussed extensively with the industrial partners and were found to be both highly insightful and useful.

The causal links between elements obtained at the workshops were then combined in a graphical network structure with the goal to obtain a generic complexity model for assembly workstations. The model consists of five clusters of variables related to complexity characterization, complexity impacts and key performance indicators (KPIs), and complexity drivers (indirect and direct), respectively. The full model is shown in Figure 4.1.

It would be too unwieldy to describe the model and its causal links in full detail. However, we will briefly treat each cluster (Figure 4.2).

Subjective complexity is perceived by the operator, and it is a combination of mental requirements and the time pressure to react on them (Figure 4.3). Mental requirements can be counted as *context switches,* which means changes in the context (parts, tools, instructions, etc.). Also, ambiguity coming from similar-looking but different parts adds to the mental workload.

When we look at the impact of the complexity elements, a dense network emerges. The main adverse impacts are

- *Loss of time* because of variant-induced workload imbalances, NVA activities linked to the number of parts, the *border of line* organization, and so on
- *Errors* that lead to scrap, rework, line stops, and a general decrease in quality level, which in turn invokes the need for additional control systems
- *Increased need for training* of the operators, because of the larger number of work instructions, the required breadth of knowledge of different models and part combinations, and their specific methods and tooling
- *Mental stress* of the operators, inducing errors, absenteeism, accidents, general frustration, and a loss of motivation and team spirit
- *Physical fatigue,* with much of the same results

These impacts will have an adverse effect on the bottom line (KPI in the model), including direct and indirect man-hour costs, investments, sales, and unproductive capacity.

Looking to the drivers' part of the causal model, we identified 11 direct drivers of complexity (Table 4.1). We will try to use (a subset of) these to quantify workstation complexity in subsequent sections.

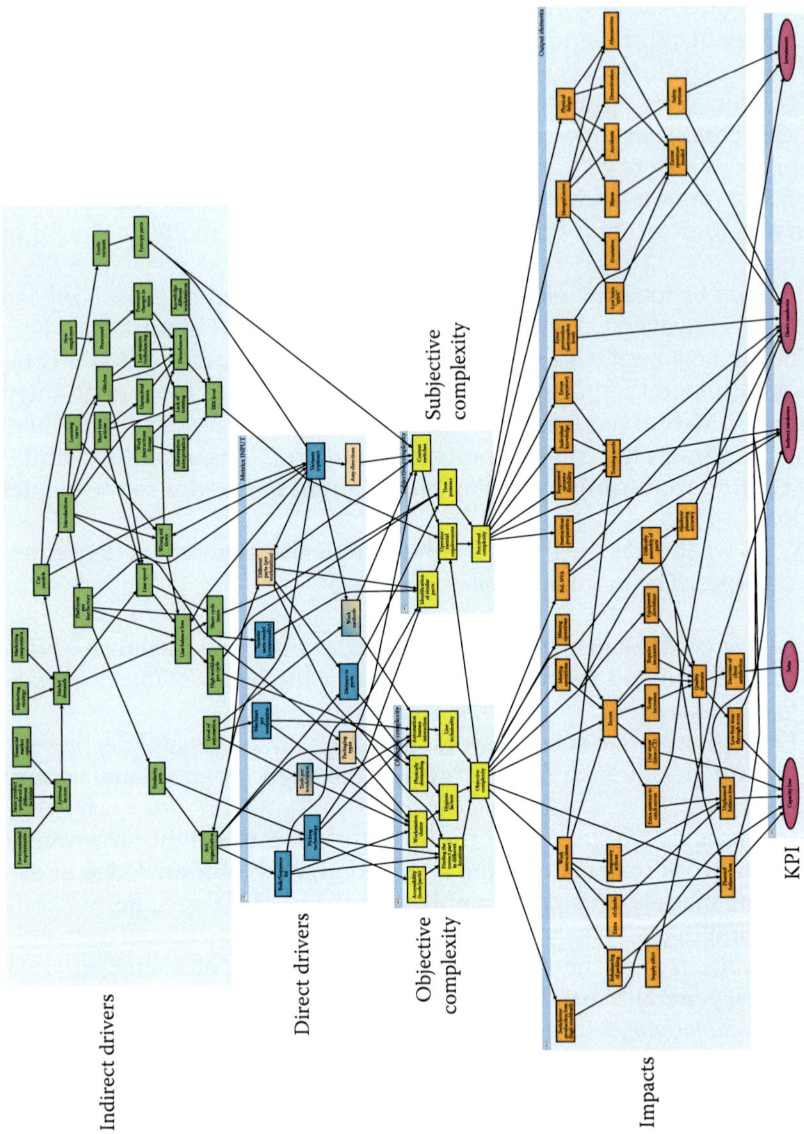

Figure 4.1 Casual model of complexity in MCM.

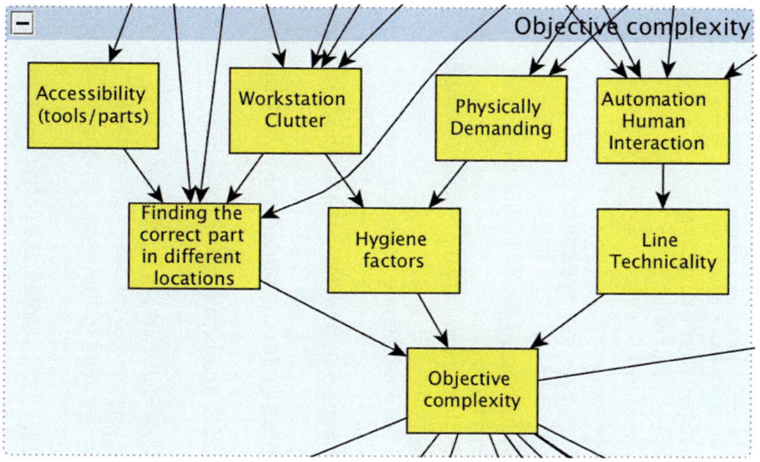

Figure 4.2 Objective complexity elements.

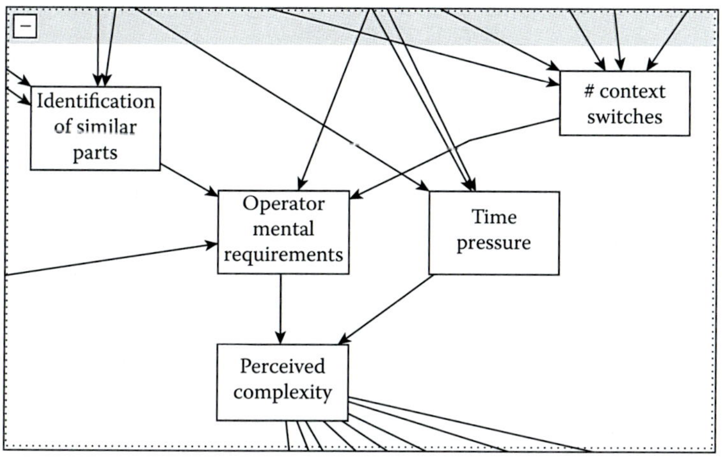

Figure 4.3 Subjective complexity elements.

These drivers are in turn influenced by a large number of elements that we grouped as *indirect drivers*. The most important ones are as follows:

- The number of *car models* and the *frequency of introduction* of new models and features to the plants is a major driver, caused by the market pressure from competitors and the context of structural overcapacity. Especially, the speed of introduction will lead to a host of adverse elements that will increase complexity, such as learning curve effects (wrong or missing work instructions, last-minute schedule changes, and glitches during ramp-up) and so on.

Table 4.1 Direct drivers of complexity

Variable name	Values or units	Description
Picking technology	(F)ixed location Pick to (S)ignal (C)omparing (M)anual	(F): Operator always takes part from the same location from bulk storage. (S): Operator picks part from location indicated by a signal (light, display). (C): Operator must compare simple information (symbols, colors). (M): Operator must read extensive information from manifest.
Bulk/sequence kit	(S)equenced kit (K)it (B)ulk	(S): Every part is in its package in the correct assembly sequence. (K): Parts are delivered in kits with the exact set for one assembly operation. (B): Parts are delivered by type in their own package.
# Packaging types	Integer number	The total number of different packaging types, a type having a specific layout. So two identical boxes with different inserts are two different types.
# Tools per workstation	Integer number	The number of tools that the operator(s) needs to handle to perform all possible assembly variants in this station, excluding automatic tools (servants).
# Machines per workstation	Integer number	Machines that perform automated tasks without operator assistance, with automatic or manual start.
# Work methods	Integer number	Each unique set of work methods the operator must master in this station. A method contains several small steps.
Distance to parts	Meters	The furthest distance between the normal operator position (or the center of the station) and the parts at the border of line.
# Variants same model	Integer number	The highest number of variants belonging to one model, among all models of which parts are assembled in this station.
# Variants in this workstation	Integer number	The total number of variant parts, combined over all models that are assembled in this workstation. So 5 types of left-hand and right-hand mirrors × 2 models = 20 variants.
# Different parts in workstation	Integer number	The total number of unique part references that are assembled in this workstation, including all variants and models that typically occur in one year.
# Assembly directions	Integer number	The number of different positions the operator must take to complete his or her workstation cycle, including repositioning of the upper body or the feet, but not small repositionings of the hands.

- The *number of options* that are offered because of marketing reasons.
- The number of different *vehicle platforms* that are built on the same line, determining the number of variants that the line (and its workstations) will have to cope with.
- Many of these elements will influence how the operator will deal with them, modeled as the skills, rules, and knowledge (SRK) level, according to the cognitive model of Rasmussen (1983).

This model reveals a complex interaction of many factors. Many of them have been researched by numerous scholars, but only a limited set have been effectively quantified. So, much research and fact finding remains to be done.

4.3 Quantifying complexity in manual assembly stations

4.3.1 A brief literature review

A major insight from the preceding discussion is that complexity is in fact a multifaceted concept with objective and subjective components. Measuring complexity is therefore a challenging task. However, as is usually done for such multifaceted concepts, attempts were made to propose indicators that may provide information of the extent to which an activity or pool of activities is complex. In this subsection, we briefly discuss a few of these indicators, mainly using entropy (more specifically, Shannon entropy).

The concept of entropy, as introduced by Shannon (1948), measures the unpredictability of information content. In the context of MCM, entropy refers to disorder or uncertainty. If operators are faced with the same models requiring the same operations, they will not experience any complexity; instead, repeating these operations will help them adopt good practices to ensure that they can execute them efficiently. It is the uncertainty of the next group of tasks to be executed by the operator that increases complexity. Entropy is therefore an appropriate indicator for complexity.

In view of the fact that complexity to some extent has to do with the overwhelming amount of information that the operator has to process while executing an activity, information-based measures were proposed. A frequently adopted approach to measure complexity is Shannon's (1948) *information entropy*. Frizelle and Suhov (2001) used as measures of complexity various (long-term) entropy rates that naturally emerge in the analysis of systems involving queues and related phenomena. Sivadasan et al. (2006) proposed a mathematical model for the operational complexity of supplier–customer systems from an information-theoretic perspective. They defined this operational complexity as the uncertainty associated with managing the time- or quantity-dynamic variations across the supplier–customer information and material flows.

Urbanic and ElMaraghy (2006) proposed a complexity model based on information content, quantity, and diversity. Information content is a relative measure of the effort needed to perform the task. Information quantity, the absolute quantity of information needed, is measured using entropy. Information diversity is the ratio of the specific information needed for a task to the total amount of information. Product complexity is determined by multiplying the product's information quantity by the sum of its content and diversity. The complexity of each process step is determined by multiplying its information quantity by the sum of the diversity ratio and the relative complexity coefficient. The relative complexity of a process step is calculated based on both cognitive and physical effort. The complexity of the whole process is the sum of the product's complexity and the sum of the complexity of all the process steps.

Zhu et al. (2008) tackled the issue of measuring variety-induced manufacturing complexity in manual mixed-model assembly lines where operators have to make choices for various assembly activities. A complexity measure they proposed is called *operator choice complexity* and is meant to quantify human performance in making choices; that is, the more choices an operator has to make, the more he or she will need additional time to process the information. Here again, the analytical form of the proposed measure is an information-theoretic entropy measure of the average randomness in the choice process. The general form of the entropy function is (Zhu et al., 2008)

$$H(X) = -\sum_{m=1}^{M} p_m \log_2 p_m \tag{4.1}$$

where:

p_m is the probability of element m occurring

M is the number of elements that can occur

The elements can be anything, from the number of parts or variants to instructions, assembly directions, tool selection, and so on. The formula illustrates the cognitive aspect that is linked with how complexity affects manufacturing systems through operator performance.

In Section 4.4.1, we show how we used this concept of entropy not only to measure the complexity of the workstation but also to rebalance the assembly line in a way that this inherent complexity may also be balanced and, hence, the adverse effects mitigated.

A very comprehensive overview and comparison of nine methods to quantify complexity has been proposed by Mattson et al. (2014). We show their main results in Table 4.2. The *complexity calculator* (CXC)

Table 4.2 Overview of nine methods to quantify complexity

Method	Aim	Type of complexity (static/dynamic)	Types of measure(s) (objective/subjective)
Operational complexity (Sivadasan et al., 2006)	To monitor and manage information and material flows	Dynamic	Objective: amount of information required to describe a state, according to flow variations, products, reasons, and variation states
Entropic measurement (Frizelle and Suhov, 2001)	To measure the rate of variety	Static and dynamic (comparison off)	Objective: probability of a state to occur according to different time measures
Manufacturing complexity index (Urbanic and ElMaraghy, 2006)	To evaluate alternatives and risk with respect to product, process, or operation task at the design stage	Dynamic	Objective: quantity, diversity, and content information in the process
Operator choice complexity (OCC) (Zhu et al., 2008)	To find causes, plan assembly sequences, and design mixed-model assembly lines	Static	Objective: average uncertainty and risk in a choice for the right tools, fixtures, parts, and procedures for variants
Knowledge and technology complexity (Meyer and Curley, 1993)	To manage software development	Dynamic	Subjective: assessment of knowledge and technological complexity
Complexity measurement (CXB) (Falck and Rosenqvist, 2012)	To support product preparation to increase productivity and decrease costs	Static and dynamic	Objective: criteria for low/high assembly complexity

(*Continued*)

Table 4.2 (Continued) Overview of nine methods to quantify complexity

Method	Aim	Type of complexity (static/dynamic)	Types of measure(s) (objective/subjective)
Robustness index (RI)	To evaluate risks and problem areas on a management/ team leader level	Dynamic	Subjective: robustness score regarding material, method, machine, and environment
Complexity calculator (CXC) (De Lima Gabriel Zeltzer et al., 2013)	To automatically assess the complexity of stations	Static and dynamic	Objective: probability that the workstation's complexity is high or low
Complexity index (CXI) (Mattsson et al., 2013)	To find problem areas at a station level	Static and dynamic	Subjective: assessment of product/variants, work content, layout, tools, and view of station

Source: Mattson et al., Comparing quantifiable methods to measure complexity in assembly, *International Journal Manufacturing Research*, 9(1), 2014, 112–130.

was developed by our team and will be treated more extensively in the remainder of this section.

As part of the same study that we described in Section 4.2, we set out to test whether the drivers, identified through the workshops, could be used to identify highly complex workstations and, if possible, even provide an objective way to quantify complexity. We also wanted to find the smallest subset of these drivers that is sufficient to provide meaningful results, in order to minimize the effort for data gathering. Since no accurate information was available regarding the "real" inherent complexity of the workstations, it was decided to ask operators and supervisors to jointly "nominate" both the most complex and the simplest workstation within their area. This subjective label was used as a benchmark throughout the study, and each industrial partner then obtained quantitative information about the driving factors for these designated workstations. The models we propose and how they were constructed is described in De Lima Gabriel Zeltzer et al. (2013).

In this way, we obtained data sets on 76 workstations from five different manufacturing locations (four in Belgium, one in Sweden), of which 41 were deemed of *LOW* complexity and 35 *HIGH*. The variables for which data was gathered are listed and explained in Table 4.1.

4.3.2 Complexity-based classification of workstations

Starting with the data on 76 workstations, we constructed a statistical model that can automatically decide whether a workstation is of LOW or HIGH complexity. Such information is very useful to identifying the workstations that warrant further analysis and appropriate methods to counter the likely effects of complexity. In statistics, the logistic or *logit* model converts a linear combination of values of characteristics into a probability belonging to either value 0 (HIGH complexity) or 1 (LOW complexity). We identified four characteristics of manual workstations (Table 4.3) that yield a very good fit for our sample, resulting in Equation 4.2.

$$P_{LOGIIT\_SAMPLE}(LOW) = \frac{e^{18.164-3.173PWL-2.326PTL-2.182ADL-0.344TWL}}{1+e^{18.164-3.173PWL-2.326PTL-2.182ADL-0.344TWL}} \tag{4.2}$$

where:

PW is the different parts in the workstation in Likert scale
PTL is the number of packaging types in Likert scale
ADL is the number of assembly directions in Likert scale
TWL is the number of tools used in the workstation in Likert scale

Table 4.3 Workstation characteristics used in the complexity quantification

Variable name	Values or units	Description
# Packaging types (PT)	Integer number	The total number of different packaging types, a type having a specific layout. So two identical boxes with different inserts are two different types.
# Tools per workstation (TW)	Integer number	The number of tools that the operator(s) needs to handle to perform all possible assembly variants in this station, excluding automatic tools (servants).
# Work methods (WM)	Integer number	Each unique set of work methods the operator must master in this station. A method contains several small steps.
# Different parts in workstation (PW)	Integer number	The total number of unique part references that are assembled in this workstation, including all variants and models that typically occur in one year.
# Assembly directions (AD)	Needing repositioning of tool/operator	The number of different positions the operator must take to complete his or her workstation cycle, including repositioning of the upper body or the feet, but not small repositionings of the hands.

Table 4.4 Coding rules to transform variables into Likert scale values

Complexity-driving variables	Likert scale coding rules (raw value)				
# Packaging types (PTL)	0	1	2–4	5–8	>8
	0	1	2	3	4
# Tools per workstation (TWL)		0–1	2–4	5–8	>8
		1	2	3	4
# Different parts in workstation (PWL)	0	1–4	5–10	11–20	>20
	0	1	2	3	4
# Assembly directions (ADL)		1	2–3	4–5	>5
		1	2	3	4

To improve the *goodness of fit*, we had to transcode some of the variables into a Likert scale (Table 4.4), inspired by MacDuffie et al. (1996). Their variable name ends with L, while those ending with R (Section 4.3.3) use the measured raw value directly.

The allocation results are shown in Figure 4.4. We can clearly distinguish the sharp transition from probability LOW to probability HIGH, yielding a distinct allocation with very few "intermediate" workstations. When we use a cutoff level of 80% to divide HIGH from LOW, we see that out of the 76 workstations, only 10 stations in the HIGH classification were (subjectively) classified as "low" by their operators (the diamond shapes in the high region) and 3 as "high" in the LOW classification.

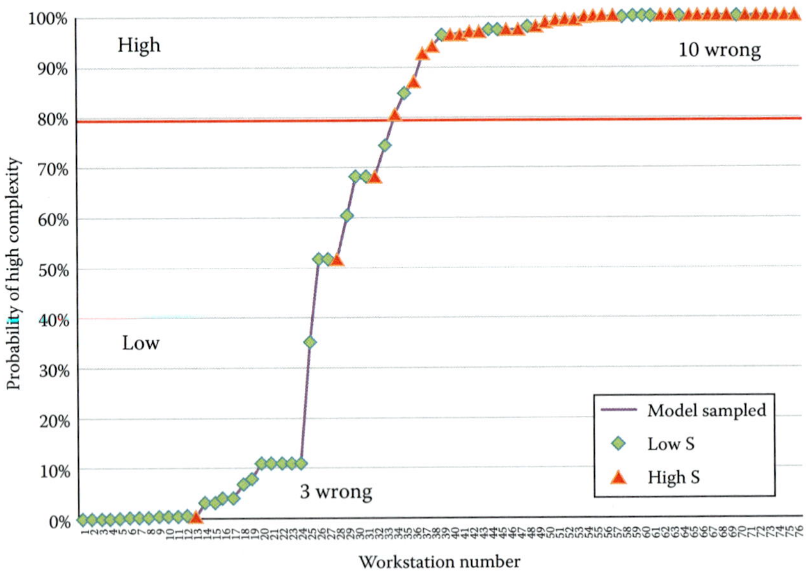

Figure 4.4 Classification results on 76 workstations.

4.3.3 Quantifying the complexity of workstations

Using the full sample of workstations, we arrived at a second logit model that yields a more gradual probability curve, which can be used to assign a complexity score to a given workstation. The model looks like this, with four variables from Table 4.1:

$$P_{LOGIT\_ALL}(LOW) = \frac{e^{6.676-1.127PTL-0.874PWL-0.243ADR-0.058WMR}}{1+e^{6.676-1.127PTL-0.874PWL-0.243ADR-0.058WMR}} \tag{4.3}$$

where:

- *PTL* is the number of packaging types in Likert scale
- *PWL* is the different parts in the workstation in Likert scale
- *ADR* is the number of assembly directions as measured directly (raw score)
- *WMR* is the number of work methods as measured directly (raw score)

The result is given in Figure 4.5. Using cutoff levels of 30% and 80%, respectively, we see that only two stations were classified differently by their operators. The overall quality of both models can be quantified using the *receiver operating characteristics* (ROC) theory (Fawcett, 2006). By differentiating the cutoff level between 0 and 1, we can generate the ROC curves of both models. Figure 4.6 shows clearly that both models are of

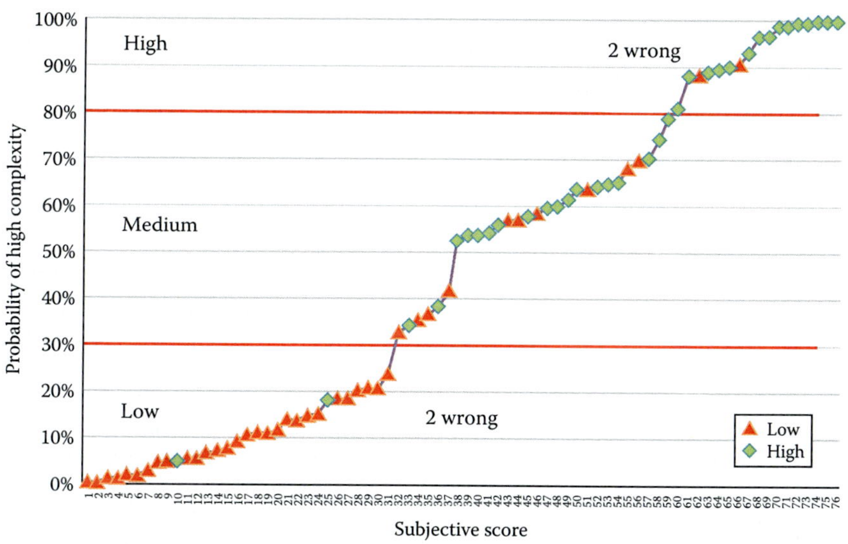

Figure 4.5 Complexity score of 76 workstations.

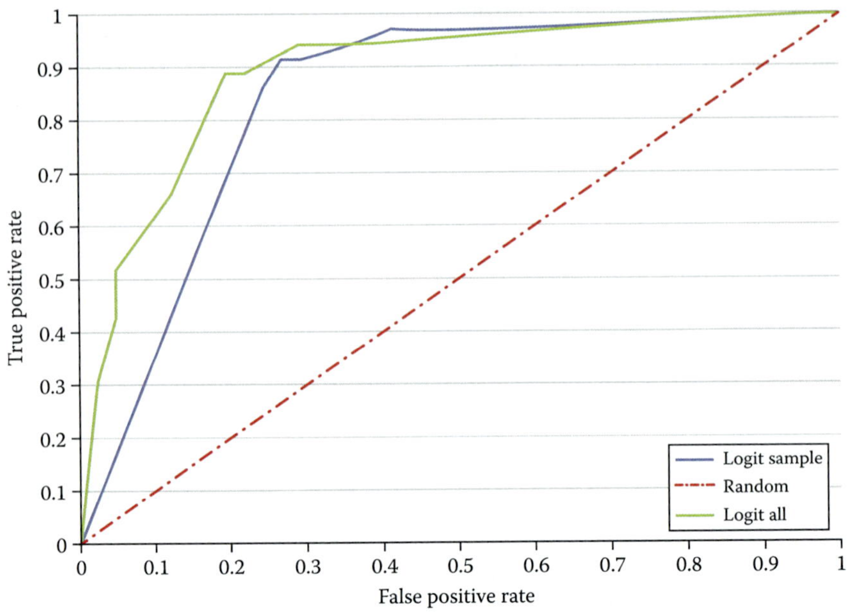

Figure 4.6 ROC curve of both complexity models.

comparable quality and equally strong. The larger the surface between the ROC curve and the 45° line (the latter indicating expected performance of a random filter), the more discriminating the model is in detecting the condition—in this case, the complexity level of the workstation.

Using both models, one can identify the high-complexity workstations, and quantify the effects of changes to the work methods and layout of the stations on its complexity score.

4.4 *Reducing the adverse impacts of complexity*

From the model in Section 4.2, we can gather a long list of the potential adverse effects of complexity. These fall broadly into two categories: effects on the manufacturing system at large and effects on the performance of the operator.

Using the complexity classification model described in Section 4.3.2, the operational characteristics of two groups of workstations was determined: 19 HIGH-complexity ones and 9 classified as LOW. The average cycle time elements are shown in Figure 4.7. We can clearly conclude that complexity increases the *balance loss* (BL) from 14% to 20% and *walking and bending* (WB) from 11% to 20%, and reduces at the same time *direct work* (DW) from 62% to 48%, or an efficiency loss of 22%! It should be noted that

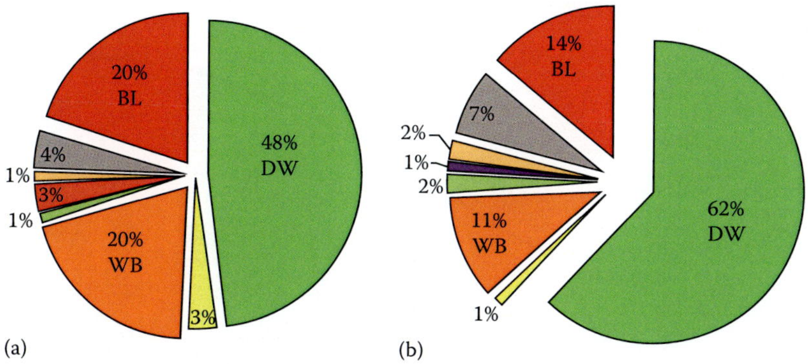

Figure 4.7 Impact of complexity on workstation efficiency. (a) 19 complex stations, (b) 19 noncomplex stations. BL, balance loss; WB, walking and bending; DW, direct work.

the increased balance loss is partly due to the strategic choice of production engineering to allot a higher common cycle time to absorb the workload peaks induced by the variants on certain workstations. If this buffer time is not sufficient, operators will fail to finish the operation or be forced to invade the working space of the next workstation. This will obviously increase the stress on the operator, which brings us to the adverse psychological effects of complexity.

To illustrate the effects of complexity on the operator's performance, we refer back to the notion of choice complexity, mentioned in Section 4.1. It is well known (Zhu et al., 2008) that the reaction time of the operator to many and diverse stimuli (which we called *context switches* in the model of Section 4.2) follows a distinct function, as shown in Figure 4.8.

So, increased complexity will increase the operator reaction time in a given work sequence within a station, further increasing the overall time pressure to the operator. A good example of this reaction time was published by Hanson et al. (2012), showing that kitting parts (preordering them in assigned locations) as opposed to bulk supply in a mixed-model assembly line at Saab considerably reduced the time to fetch the parts (Figure 4.9). Also, the manner in which part identification was conveyed (printed manifest vs. signal lights) had an impact. Since fetch time includes information reaction time as well as the physical grasping of the part, it illustrates how cognitive load influences operator performance.

As already introduced at the beginning of this chapter, the main effect of complexity is an increased cognitive load on the operator. In our causal model, we identified many aspects of this so-called subjective complexity. To avoid going extensively into a psychological analysis of

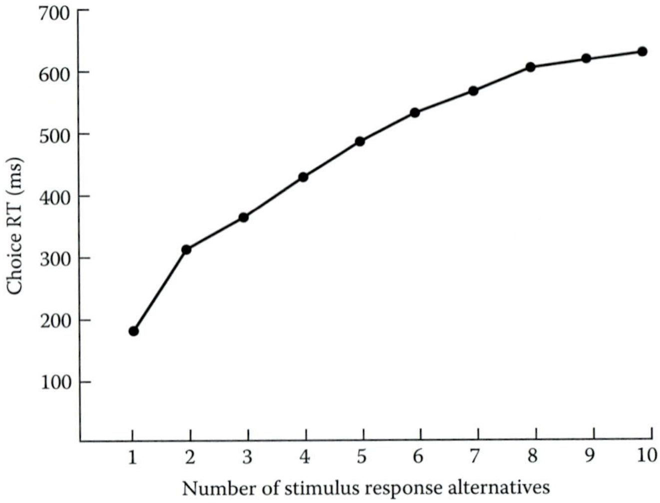

Figure 4.8 Operator reaction time to choice complexity.

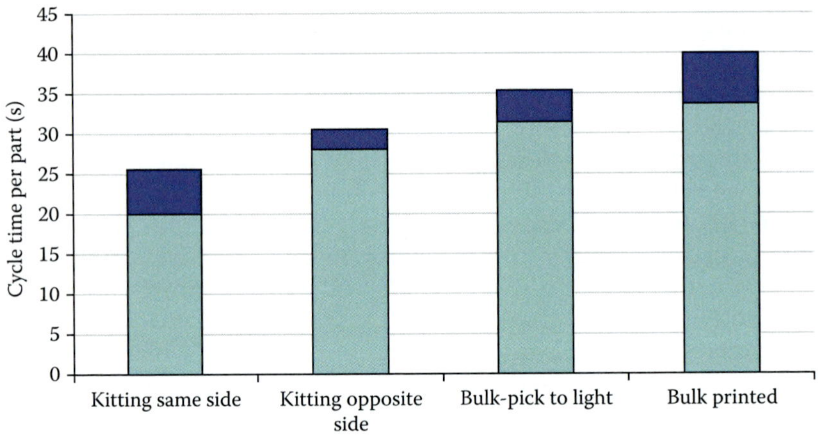

Figure 4.9 Effect of complexity on operator reaction time (after Hanson et al., 2012); dark blue regions are +/−1 standard deviation around the mean.

the cause and effect of complexity, we focus on the Rasmussen model (Table 4.5). The main effect of complexity is to present the operator with a larger proportion of level 2 (rules) and level 3 (knowledge) problems, to the detriment of the level 1 skills-based tasks for which he or she is trained, and on which most of the allotted production times are based.

Table 4.5 Complexity-induced cognitive load and mitigating actions

Level	Rasmussen classification of operator reaction required	Example	Mitigating technology
1	Skill based	Repetitive assembly of different part sets on different variant models using the right tools and applying the right procedure (torque, visual control, etc.), incurring varying workload per cycle	Online tool sensoring and active tags for error control, increased balance loss or peak shaving sequencing
2	Rule based	Assembly of rare parts from infrequent variants requiring different tools and/or procedures and access to instructions	Wearable augmented reality devices
3	Knowledge based	Unique parts for which standard instructions are not available or not up to date, requiring "same as but" behavior and supervisor assistance within normal cycle time	Context-sensitive information system generating on-the-fly step-by-step instructions, empowered operators

4.4.1 Line balancing to reduce work overload and smoothen complexity

The assembly platform on which the various automotive models are assembled, requiring different operating times at each workstation, is known in the literature as a mixed-model assembly line (Figure 4.10). In a mixed-model assembly line, more than one model of the same general product is intermixed and assembled on the same line. The amount of work required to assemble units can vary from model to model, creating an uneven flow of work along the line. When the cycle time is fixed and new models are added to the portfolio of the models on the line, work overload may occasionally or regularly occur in some workstations. Increasing the cycle time for the whole line to moderate overload occurrences is usually not acceptable in a competitive environment such as the automotive industry. Instead, rebalancing the line

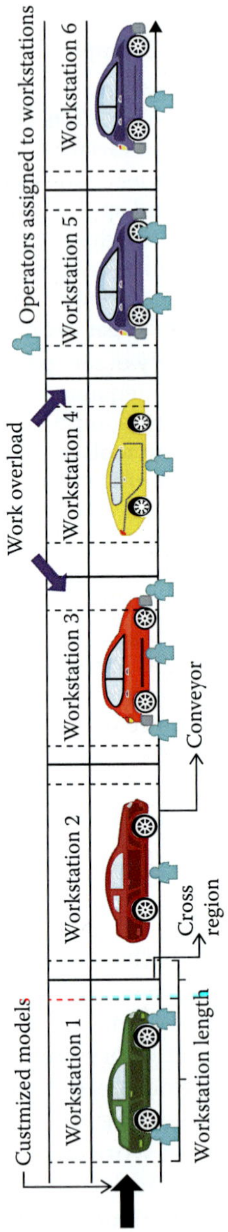

Figure 4.10 Example of a mixed-model assembly line. (From Zeltzer Luiza, PhD dissertation, Ghent University, Belgium, 2016.)

occasionally and sequencing the builds during each cycle to minimize the frequency and magnitude of overloads is performed by the manager of these lines.

In this section, we present and discuss some optimization models and techniques to minimize overload via intermittent line rebalancing and regular sequencing of the planned vehicle mix (builds). We then examine how an entropy-based measure of complexity can be used to reduce complexity while rebalancing the line. Some of these models are tested on some real data sets obtained from our partners in the automotive industry and some of their suppliers. The major results are also presented and discussed.

For a clear presentation of the optimization models and line-balancing approach, we define the following relevant parameters:

We let M be the set of car models to be assembled on the mixed-model line, indexed by m. For each model type m, let d_m be the demand proportion that the model type m represents in the total model-mix demand assembled on the line (note that $\Sigma_{m\in M} d_m = 1$). Let K be the set of workstations making up the assembly line, indexed by k. Each workstation k has a length L^k_{maxk} measured in time units (Figure 4.11).

We let J be the set of all tasks required to assemble the various models in M, indexed by j. Each task j has a set *Pred (j)* of direct predecessor tasks that must be performed before j. We use O to designate the set of operators on the line. We also let J_k be the set of workstations k on which task j can be performed and J_l be the set of operators l qualified to perform task j. We also introduce the parameter $O^k{}_{max}$, which represents the maximum number of operators that can be assigned to a workstation k. This particular possibility is usually not taken into account in existing line-balancing procedures. Finally, we let c denote the targeted cycle time of the line and t_{jm} the processing time of task j for the model m.

Given these parameters, the so-called mixed-model assembly line balancing problem (MMALBP), which is an optimization problem to optimally design the assembly line, consists of finding an optimal assignment of all the tasks in J to the workstations in K that satisfies the various constraints resulting from task precedence relationships together with the qualified workstations and operators restrictions, and this while

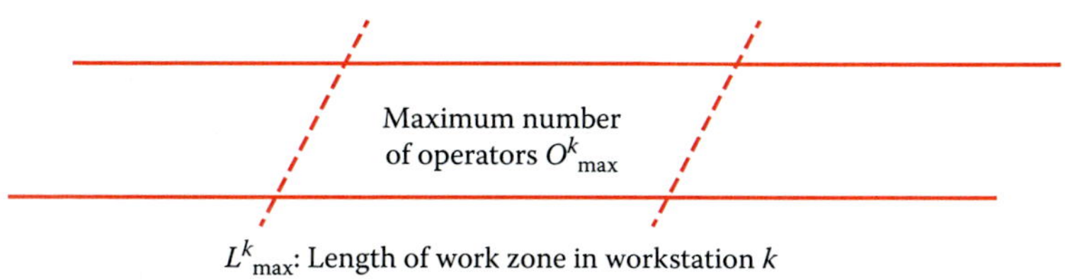

Figure 4.11 Workstation's major parameters.

minimizing the work overload (and hence the balance loss) resulting at each workstation when some car models are assembled.

The procedure and the underlying optimization tool to achieve such a line balance is summarized in the following section.

4.4.1.1 Phase 1: Mixed-model line-balancing optimization

- *Initialization step*:
  - Select a *supermodel* (to reduce the problem to a single product case), Type I, II, F, or E, and then select a heuristic algorithm to determine an initial line balance.
  - Determine the corresponding workstations and operator workloads and all work overloads across all models.
- *Improvement step*:
  - Improve the current workload balance (eliminate or reduce all overloads) using the selected heuristic algorithm until the stopping criteria is satisfied (minimum number and magnitude of overloads).
- *Analysis and visualization step*:
  - Visualize the final line balance and compute station and operator workloads and overloads and all relevant statistics, such as the number of overloads and spreads per station and operator.

For the sake of completeness, we also recall here the widely used definitions of the various basic line-balancing models: in Type I the objective is to minimize the number of workstations given the cycle time; in Type II the objective is to minimize the cycle time for a given number of workstations; in Type F the objective is to determine a feasible line balance, given the number of workstations and the cycle time; finally, in Type E the objective is to minimize both the cycle time and the number of stations.

Note at this point that the use of a supermodel to reduce the problem to a simple line-balancing problem is not the only alternative. Technically, one can solve the problem taking the different models explicitly into account. However, the use of a supermodel has an advantage in terms of computational time and understanding by the users in the field if they wish to use the line-balancing tools as a black box.

Figure 4.12a and b show a typical summary of the results that must be generated after the analysis and visualization step of the procedure. These summary figures show the number of work overload occurrences, the spread per workstation, and the models to be handled carefully. These important performance measures will be exploited during the sequencing phase.

The initialization step requires that a supermodel be defined. A supermodel is characterized by the processing time of each task j, denoted as $\hat{t}_j = f(t_j^1,\ldots,t_j^m,\ldots,t_j^M,d_1,\ldots,d_m,\ldots,d_M)$, and is a function of the model

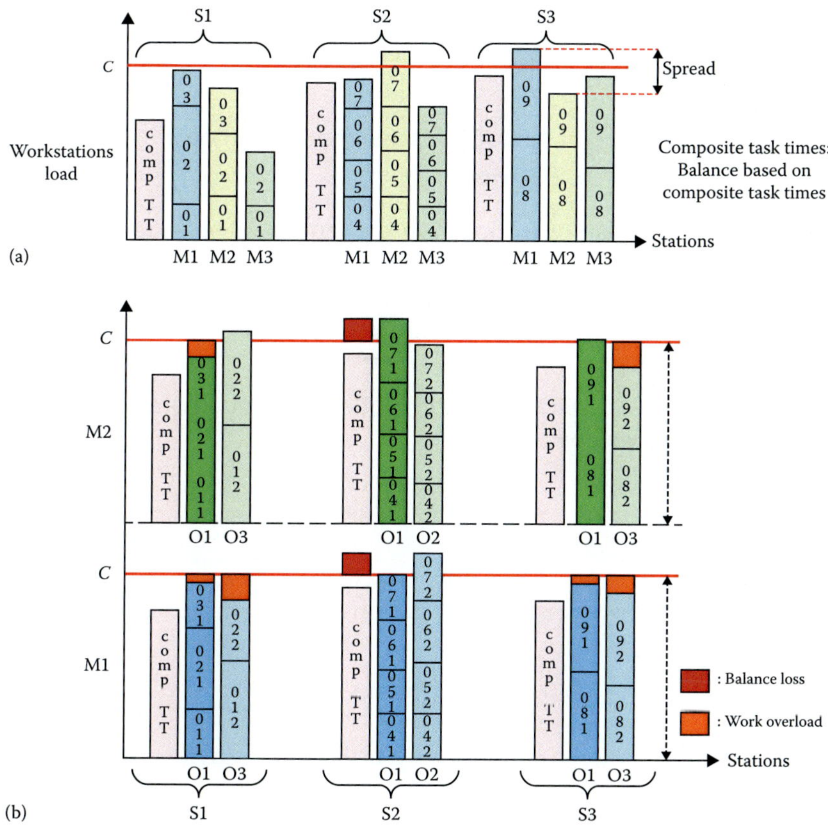

Figure 4.12 (a) Workstation's major statistics: maximum spread per station. (b) Balance loss vs. overload statistics per operator, workstation, and model.

processing times and their relative demand ratios. There are many alternative ways to define this function. We propose examples of such a function in the following:

- Weighted average times: $\hat{t}_j = \sum_{m=1}^{M} d_m t_j^m$
- Specific model m*: $\hat{t}_j = t_j^{m^*}$
- Average times: $\hat{t}_j = \sum_{m=1}^{M} t_j^m / M$
- Maximum task time: $\hat{t}_j = Max_{m=1,\ldots,M} t_j^m$
- Minimum task time: $\hat{t}_j = Min_{m=1,\ldots,M} t_j^m$
- Most occurring task time: $\hat{t}_j = MostFrequent_{m=1,\ldots,M} t_j^m$

Once a supermodel is selected, the following decision variable can now be used within an optimization model to determine a task assignment that balances the resulting workload across the workstations.

$$x_{jk}^{l} = \begin{cases} 1, \text{ if task } j \text{ is assigned to station } k \text{ and operator } l \\ 0, \text{ otherwise} \end{cases} \tag{4.4}$$

$$y_{lk} = \begin{cases} 1, \text{ if operator } l \text{ is assigned to station } k \\ 0, \text{ otherwise} \end{cases} \tag{4.5}$$

The total workload of workstation k when model m is loaded is given by

$$\tau_{k}^{m} = \sum_{j \in J} t_{j}^{m} \sum_{l \in O} x_{jk}^{l} \tag{4.6}$$

The workload of operator l when model m is loaded on workstation k is given the same way by

$$\tau_{lk}^{m} = \sum_{j \in J} t_{j}^{m} x_{jk}^{l} + CL_{k} \tag{4.7}$$

where CL_k is the *complexity allowance* on workstation k.

The average processing time at station k is determined as

$$T_{k} = \sum_{m=1}^{M} d_{m} \max\left\{\tau_{lk}^{m} : l \in O\right\} \tag{4.8}$$

The average station time assuming tasks can be divided at will:

$$\bar{T} = \sum_{k=1}^{K_{\max}} T_{k} / K_{\max} \tag{4.9}$$

Average processing time of model m over all stations:

$$T_{m} = \sum_{k=1}^{K_{\max}} \max\left\{\tau_{lk}^{m} : l \in O\right\} / K_{\max} \tag{4.10}$$

Typical objective functions that can be optimized in this context are given by Equations 4.11 and 4.12, involving maximal deviations of each operator's workload for each model and workstation from the cycle time or average station time

$$\max_{k=1 \ldots K_{\max}, l=1,\ldots,L_{\max}(k), m=1 \ldots M}\left\{\left|\tau_{lk}^{m} - (C_{\max}, T_{m} \text{ or } \bar{T})\right|\right\} \tag{4.11}$$

or

$$\sum_{k=1}^{K_{\max}} \sum_{l=1}^{L_{\max}^{k}} \sum_{m=1}^{M} \max\left\{0, \tau_{lk}^{m} - (C_{\max}, T_{m} \text{ or } \bar{T})\right\} \tag{4.12}$$

where $L^k_{\max}$ is the maximum number of operators working in parallel that can be assigned to workstation k.

The constraints to be satisfied are

- Task assignment (1):

$$\sum_{k\in K_j}\sum_{l\in O_j} x^l_{jk} = 1, \text{ for all } j \in J \tag{4.13}$$

- Precedence (2):

$$\sum_{l\in O_j} x^l_{jk} - \sum_{h\in K_i, h\le k}\sum_{l\in O_i} x^l_{ih} \le 0, \text{ for all } j \in J, \text{ and } i \in P_j \tag{4.14}$$

- Operator assignment (3):
    - Required if each operator can only be assigned only at most one workstation:

$$\sum_{k\in W} y_{lk} \le 1, \text{ for all } l \in O \tag{4.15}$$

    - Number of operators working in parallel on workstation k:

$$\sum_{l\in O} y_{lk} \le L^k_{\max}, \text{ for all } k \in W \tag{4.16}$$

where $L^k_{\max}$ is the maximum number of parallel operators on station k.

- Station's and operator's workload (4):

$$\sum_{j:l\in O_j} x^l_{jk} - |J_l \cap J_k| y_{lk} \le 0, \text{ for all } l \in O \text{ and } k \in W \tag{4.17}$$

When using a supermodel:

$$\sum_{j:l\in O_j} \hat{t}_j x^l_{jk} \le (C_{\max} + CL_k), \text{ for all } l \in O \text{ and } k \in W \tag{4.18}$$

To directly control maximum deviations:

$$\sum_{j:l\in O_j} t_{jm} x^l_{jk} \le (C_{\max} + CL_k) + \Delta^l_{mk}, \text{ for all } l \in O, m \in M \text{ and } k \in W \tag{4.19}$$

where:

J_l is the set of all tasks that can be performed by operator l

J_k is the set of all tasks that can be performed on station k

Δ^l_{mk} is the overload caused by model m to operator l's workload

This optimization model can be used in many ways.

- In cases where the number of workstations and the cycle time are given, the optimization model can be used to determine the best feasible task and operator assignments, balancing the workloads and minimizing overloads.
- If tasks are already assigned, then by fixing x variables the model turns into a generalized assignment optimization model, which optimizes operator assignments, balances their workloads, and minimizes overloads.
- In the case of one operator per station, the optimization model turns into the classical line-balancing model.

4.4.1.2 *Phase 2: Sequencing and work overload optimization*

Figure 4.13a and b show that the sequence in which the models are assembled might negatively impact the productivity of the line. If the work overload at a workstation is not resorbed at the subsequent ones, the productivity of the line will be negatively impacted. After balancing the line to reduce as much as possible the number and magnitude of structural work overloads, sequencing contributes in helping to resorb some of these structural overloads.

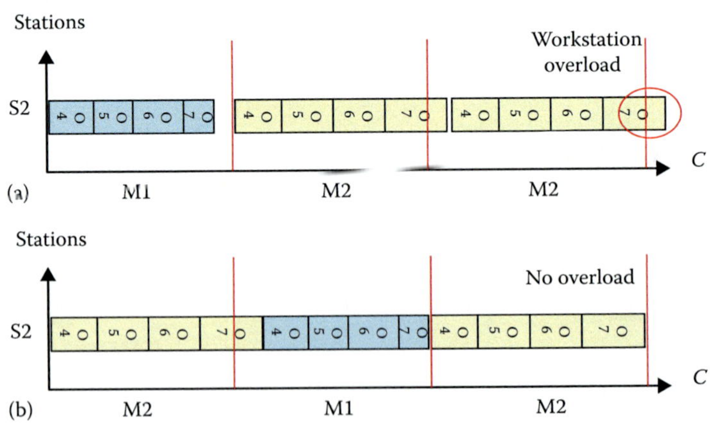

Figure 4.13 (a) If the builds are not sequenced, overload occurs. (b) If the builds are sequenced, no overload occurs.

4.4.1.3 Analysis of a prototype example

To illustrate the model approaches discussed in this section, we consider a data set consisting of nine workstations, nine models, nine operators, and 109 tasks. First, we present and compare the results of the mixed-model assembly line-balancing model. Then, we analyze the impact of sequencing on the work overload at each work station.

Figure 4.14 shows the results of the current balance. One can observe that some overloads occur because some new models were added to the portfolio of the models assembled on the line without rebalancing. In workstation 7 for instance, model 2 produces a high underload, whereas model 3 produces a high overload.

The use of the optimization produces alternative assignments of tasks and consequently different loads at the workstations. Figure 4.15 shows

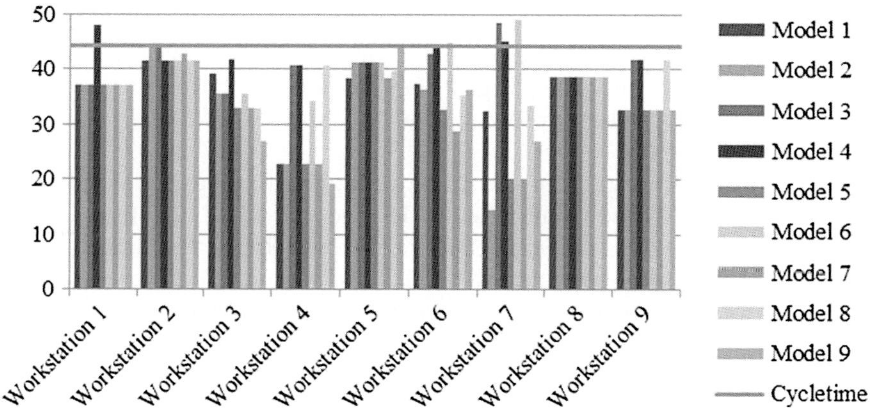

Figure 4.14 Results of the current line balance.

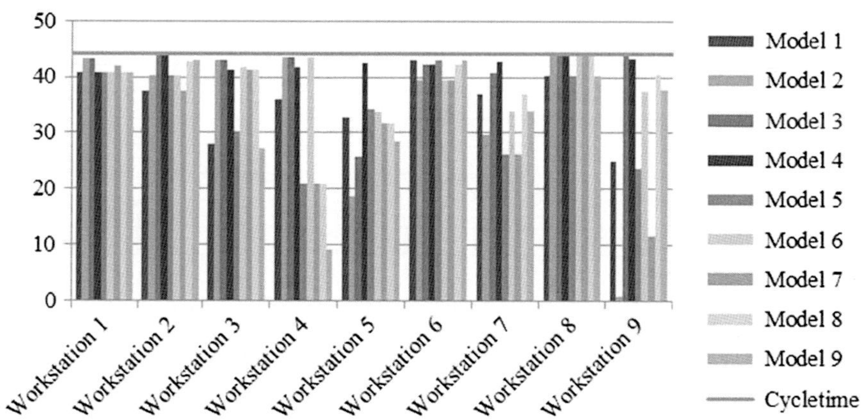

Figure 4.15 Results of the optimized line balance.

the results of the optimal line balance obtained via the proposed optimization model. One can see that there are no overloads left, and even the sequencing is not necessary in this case.

To conclude this section, we now discuss how complexity can be taken into account when developing a line-balancing model. After a first balance is developed, we can compute the entropy at each workstation in the usual way.

$$Complexity_k = -\sum_{j \in S_k} p_j \log(p_j) \tag{4.20}$$

where p_j is the probability of occurrence of task j given the models that are assembled on the workstation, and S_k is, of course, the set of tasks assigned to workstation k. We can then report these values for each workstation on a graphic, as shown in Figure 4.16. In this figure, the red line labeled "Original workstation entropy" shows the entropy at workstations 6 and 7 are high. These two workstations are, by the way, among those that were revealed as complex by the analysis in Sections 4.2 and 4.3.

After rebalancing, the green line in Figure 4.16 gives the entropy at the workstation. These values are now leveled and are almost the same at each workstation. One can expect now that this rebalancing has not only reduced overload but also helped reduce the complexity experienced by the operators. The next section elaborates on this important issue.

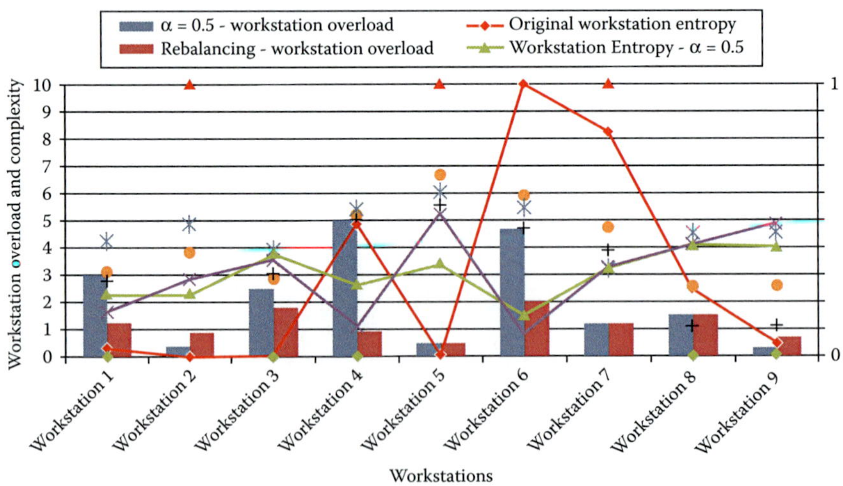

Figure 4.16 Summary of manufacturing complexity and line-balancing analyses. (From Zeltzer Luiza, PhD dissertation, Ghent University, Belgium, 2016.)

4.4.2 Operator information systems to reduce mental workload

Careful inspection of Figures 4.2 through 4.4 reveals that, despite advances in automation and production equipment technology, information to and from the operator will take the leading role in mitigating the effects of complexity. Whenever the context surpasses level 1 behavior (Table 4.5), information technology can step in. In a first attempt to trigger level 2 actions, information systems can retrieve the procedure with the best fit to the observed context and present the operator with the right tasks. If no suitable predefined procedure exists, the system will need operator instructions to be able to present the operator with information, as well as record the context for possible offline treatment by engineering (Claeys et al., 2015).

Selecting the right amount of information so as not to overburden the operator (which would increase his or her cognitive load rather than subdue it) and presenting it in a nonobtrusive manner so as not to interfere with production tasks remains a big challenge. The fast-evolving technology of wearable devices is emerging as a very likely solution. Already, large companies such as BMW (Woollaston, 2014), but also small ones such as industrial bakeries (Vandorpe, 2014), use augmented reality goggles to support operators in complex and error-prone tasks (Figure 4.17).

A second important reason to adopt these bidirectional operator information systems is to allow operators to use their experience and knowledge and capture them for later use. Operators should also be able to annotate information and flag parts that are inadequate or plainly wrong. This approach could be a rich source of improvement targets for offline treatment by production engineering, and also enable semiautomatic generation of correct operating instructions for training purposes and to secure improvements.

Figure 4.17 Augmented reality on the shop floor. (Courtesy of VUZIX.)

One important aspect of such information systems that still needs to be developed is context gauging. The system should at all times be able to infer from both the physical appearance of the workstation (the objective complexity) and the cognitive situation of the operator (the subjective complexity) what information is needed and what automated actions are possible. To make this happen, research is needed to gain a better understanding of the interaction and communication between the human assisted by technology and the production process. This will require accurate capturing of the cognitive status of the operator (stress levels, cognitive level of information processing, etc.) as well as his or her relevant physical actions. These actions can be conscious and deliberate but also more unconscious and emotional. We can distinguish different types of physical actions, such as (1) the manipulation of the work piece (using tools or hands), (2) the control of a machine through an interface (using levers, buttons, joysticks), (3) the maintenance of the workplace (setup, repair, clean), and so on. Different cognitive actions such as (1) work instructions, (2) rapid problem-solving, and (3) emotional states (e.g., stress, fatigue, satisfaction) should also be studied.

4.5 Conclusions

In this chapter, we have explored the main elements of complexity and how it impacts the performance of manual assembly workers in a high-variety setting. Through a causal model, we have shed more light on the different drivers of complexity, the mechanisms with which complexity interferes with manufacturing performance and the cognitive performance of the workforce, and the different consequences. We also proposed a quantitative approach to classifying workstations as high or low complexity and subsequently a gradual measure of complexity. By studying the differences between the two kinds of workstations, we described objective degradations of workstation efficiency, as well as the heightening of cognitive load, according to a mental model. Finally, we included two approaches to mitigating complexity in manufacturing. A specific line-balancing algorithm was proposed to reduce workload peaks in high-complexity workstations (allowing a reduction in overall balance loss and complexity), and a technology-driven approach to operator information systems was described to mitigate stress related to cognitive overload.

The coming years will see both focused research into this domain of human-centered manufacturing systems and a wave of practical experiences from early adopters that encompass all types and sizes of companies, as was illustrated.

References

Claeys, A., Hoedt, S., Soete, N., Cottyn, J., and Van Landeghem, H. (2015). Framework for evaluating cognitive support in mixed model assembly systems. IFAC Symposium on Information Control in Manufacturing (INCOM 2015). *IFAC-PapersOnLine* 48(3), 980–985.

De Lima Gabriel Zeltzer, L., Limère, L., Van Landeghem, H., Aghezzaf, E. H., and Stahre, J. (2013). Measuring complexity in mixed-model assembly workstations. *International Journal of Production Research*, 51(15), 4630–4643.

Deshmukh, A. V., Talavage, J. J., and Barash, M. M. (1998). Complexity in manufacturing systems, Part 1: Analysis of static complexity. *IIE Transactions*, 30(7), 645–655.

ElMaraghy, W. H., ElMaraghy, H., Tomiyama, T., and Monostori, L. (2012). Complexity in engineering design and manufacturing. *CIRP Annals: Manufacturing Technology*, Retrieved November 16, 2015 from http://dx.doi.org/10.1016/j.cirp.2012.05.001.

ElMaraghy, W. H., and Urbanic, R. J. (2004). Assessment of manufacturing operational complexity. *CIRP Annals*, 53(1), 401–406.

ElMaraghy, W. H., and Urbanic, R. J. (2003). Modeling of manufacturing system complexity. *CIRP Annals*, 52(1), 363–366.

Falck, A.-C., and Rosenqvist, M. (2012). Relationship between complexity in manual assembly work, ergonomics and assembly quality. *Ergonomics for Sustainability and Growth*, NES 2012, Stockholm, Sweden.

Fawcett, T. (2006). An introduction to ROC analysis. *Pattern Recognition Letters*, 27, 861–874.

Fisher, M. L., and Ittner, C. D. (1999). The impact of product variety on automobile assembly operations: Empirical evidence and simulation analysis. *Management Science*, 45, 771–786.

Fisher, M. L., Jain, A., and MacDuffie, J. P. (1995). Strategies for product variety: Lessons from the auto industry. In B. Kogut and E. Bowman, Eds. *Redesigning the Firm*, pp. 116–154. Oxford: Oxford University Press.

Frizelle, G. (1996). Getting the measure of complexity. *Manufacturing Engineer*, 75(6), 268–270.

Frizelle, G., and Suhov, Y. M. (2001). An entropic measurement of queuing behaviour in a class of manufacturing operations. *Proceedings of Royal Society, London*, 457, 1579–1601.

Fujimoto, H., and Ahmed, A. (2001). Entropic evaluation of assemblability in concurrent approach to assembly planning. *Proceedings of the IEEE International Symposium on Assembly and Task Planning*, 306–311.

Fujimoto, H., Ahmed, A., Iida, Y., and Hanai, M. (2003). Assembly process design for managing complexities because of product varieties. *International Journal of Flexible Manufacturing Systems*, 15, 283–307.

Hanson, R., and Medbo, L. (2012). Kitting and time efficiency in manual assembly. *International Journal of Production Research*, 50 (4), 1115–1125.

Hu, S. J., Zhu, X., Wang, H., and Kkoren, Y. (2008). Product variety and manufacturing complexity in assembly systems and supply chain. *CIRP Annals: Manufacturing Technology*, 57, 45–48.

MacDuffie, J. P., Sethuraman, K., and Fisher, M. L. (1996). Product variety and manufacturing performance: Evidence from the international automotive assembly plant study. *Management Science*, 42(3), 350–369.

Mattsson, S., Karlsson, M., Gullander, P., et al. (2014). Comparing quantifiable methods to measure complexity in assembly. *International Journal Manufacturing Research*, 9(1), 112–130.

Mattsson, S., Gullander, P., Harlina, U., Bäckstrand, G., Fastha, Å., and Davidsson, A. (2012). Testing Complexity Index: A method for measuring perceived production complexity. *Procedia CIRP* 3, 394–399.

Meyer, M. H., and Curley, K. F. (1993). The impact of knowledge and technology complexity on decision making software development. *Expert Systems with Applications*, 8(1), 111–134.

Parasuraman, R., Sheridan, T., and Wickens, C. (2000). A model for types and levels of human interaction with automation. *IEEE Transactions on Systems, Man, and Cybernetics, Part A: Systems and Humans*, 30(3), 286–297.

Rasmussen, J. (1983). Skills, rules, and knowledge: Signals, signs, and symbols, and other distinctions in human performance models. *IEEE Transactions on Systems, Man and Cybernetics*, SMC-13 (No. 3, May/June), 257–266.

Rodríguez-Toro, C., Jared, G., and Swift, K. (2004). Product-development complexity metrics: A framework for proactive-DFA implementation. *Proceedings of DESIGN 2004, the 8th International Design Conference*, Dubrovnik. Bristol: The Design Society, pp. 483–490.

Samy, S. N., and ElMaraghy, H. (2010). A model for measuring products assembly complexity. *International Journal of Computer Integrated Manufacturing*, 23(11), 1015–1027.

Schuh, G., Monostori, L., Csájib, B. C., and Döringa, S. (2008). Complexity-based modeling of reconfigurable collaborations in production industry. *CIRP Annals*, 57(1), 445–450.

Shannon, C. E. (1948). A mathematical theory of communication. *Bell System Technical Journal*, 27(3), 379–423.

Sivadasan, S., Efstathiou, J., Calinescu, A., and Huatuco, L. H. (2006). Advances on measuring the operational complexity of supplier-customer systems. *European Journal of Operational Research*, 171(1), 208–226.

Urbanic, R. J., and ElMaraghy, W. H. (2006). Modeling of manufacturing process complexity. In H. A. ElMaraghy and W. H. ElMaraghy, Eds. *Advances in Design*. Springer Series in Advanced Manufacturing, ISBN 978-1-84628-004-7, pp. 425–436.

Vandorpe, A. (2014). Biobakkerij pioniert met smart glasses (Bio-bakery pioneers with smart glasses), *De Tijd*, March 3, 2014. Retrieved December 1, 2015 from http://www.tijd.be/ondernemen.

Woollaston, V. (2014). The end of the mechanic? *Daily Mail*, January 21, 2014. Retrieved December 1, 2015 from http://www.dailymail.co.uk/sciencetech/article-2543395.

Zeltzer, L. (2016). Analysing and levelling manufacturing complexity in mixed-model assembly lines. PhD dissertation, Faculty of Engineering and Architecture, Ghent University, Belgium.

Zhu, X. (2009). Modeling product variety induced manufacturing complexity for assembly systems design. Dissertation, University of Michigan, Ann Arbor, MA, p. 89. Retrieved November 16, 2015 from http://hdl.handle.net/2027.42/62265.

Zhu, X., Hu, J. S., Koren, Y., and Marin, S. P. (2008). Modeling of manufacturing complexity in mixed-model assembly lines. *Journal of Manufacturing Science and Engineering*, 130, 051013-051013-10. doi:10.1115/1.2953076.

chapter five

Modeling of assembly supply chain structures

Vladimir Modrak and Slavomir Bednar

Contents

ABSTRACT

Assembly supply chain (ASC) structures and their modeling can be effectively used when planning assembly sequences in terms of mass customized manufacturing. Graph theory is commonly applied for this purpose. Generally, assembly sequence planning helps layout designers, among others, to increase productivity and to decrease complexity. On the other hand, supply chains are required to be highly flexible to obtain short periods of order realization and high product diversity. Especially, reducing network complexity and finding alternative ASCs has recently been at the center of managers' attentions, as ASC systems are becoming increasingly complex. This chapter provides a framework to model alternative ASC structures, which are subsequently analyzed using selected structural complexity indicators. Later, we outline the initial components of mass customized assembly and present the modeling of available product configurations on the hypothetical model of a labeled graph.

5.1 Introduction

The gradual transformation of companies toward mass customization (MC) is pulled by the growing demand for tailor-made mass-produced products and pushed by the rapid development of modern supporting technologies such as information technologies, additive manufacturing, fifth-generation mobile networks, identification technologies, and others. This gradual development causes companies to take on different forms of MC. In this context, Stump and Badurdeen [1] differ between low-level MC and high-level MC. In the second case, it is required that manufacturing systems are highly flexible and that manufacturing planning and control are more complex. As a consequence, requirements on process modularity are of higher importance. Therefore, companies with a higher degree of customization might focus on reducing the complexity of their production processes, since complexity problems not only affect production processes but also managerial processes [2]. Although this objective is quite clear, the way to achieve it is not. The identification of complexity metrics is the first precondition for solving the problem. Prior to this task, it would be helpful to have a basic understanding of what factors affect the manufacturing complexity. A typical feature for MC is that products consist of several modules and each module can have a certain number of variants. Combinations of these variants contribute to high product variety, which triggers high manufacturing complexity [3]. The assembly of the modules creates a network of interconnected workstations that are frequently characterized as mixed-model assembly systems. At each station, selected components are assembled onto the partially finished product. The end product is then finalized at the last station. According to Koren et al. [4], a configuration of assembly stations has a notable impact on the performance of manufacturing systems. Wang [5] therefore adds that "it is necessary to take into account the effect of system configuration when studying the variety-induced manufacturing complexity and its impact on the performance of mixed-model assembly systems."

The chapter is divided into two parts. The first part proposes a framework for modeling alternative ASC structures and gives a brief description of possible complexity indicators, with the aim to measure the structural complexity of alternative ASCs. In the second part of the chapter, an approach to determining all possible product configurations is outlined. The quantification of all possible product configurations and variants will be treated in detail in Chapter 6.

5.2 Modeling of assembly supply chain structures

Modeling of ASC structures is especially useful when the structures obtained as a result can be utilized for practical guidance and applied in conjunction with a selection of optimal assembly process structures. Generally, assembly sequence planning helps layout designers, among others, to increase productivity and reduce costs and complexity. In particular, network complexity reduction has recently been at the center of managers' attention as ASC systems are becoming increasingly complex. On the other hand, supply chains are required to be highly flexible to obtain short delivery times.

5.2.1 Generating all possible assembly supply chain structures

Graph theory is commonly applied for this purpose. ACSs can be represented by directed tree graphs, in which each node in the chain has at most one successor but may have any number of predecessors. Such supply chain structures are convergent and are divided into two basic types: modular and nonmodular. In the modular structure, the intermediate subassemblers are understood as assembly modules, while the nonmodular structure consists only of original suppliers and a final assembler (the root node). Steps to identify optimal ASC structures are clearly specified by Zhu et al. [6] as follows:

1. Generation of all possible supply chain structures
2. Quantification of topological complexity values for each possible configuration
3. Selection of the optimal supply chain configurations

The same authors outlined the way forward to model possible supply chain structures depending on the number of original suppliers *i*. For example, if the supply chain has between 2 and 5 original suppliers, then we obtain 1, 2, 5, and 12 different ASC networks, respectively, as can be seen in Figure 5.1.

Analogically, it is possible to generate topological structures of ASC networks for higher numbers of original suppliers [7]. For this purpose, it is useful to establish a framework for creating so-called topological classes of ASCs for both nonmodular and modular ASC networks based on the number of initial nodes (original suppliers) *i*, respecting the following rules:

1. ASC structures with an identical number of initial nodes are grouped into so-called topological classes categorized by the number of initial nodes.

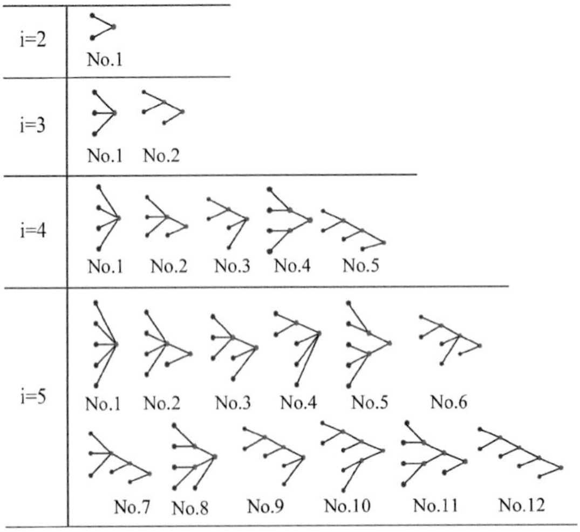

Figure 5.1 All possible ASC networks based on the number of original suppliers.

2. The initial nodes i in ASC structures are located on tiers t_l ($l = 0,\ldots, m$). The tiers are ordered from right to left.
3. A final assembly workstation is situated in tier t_0, while it is assumed to model ASCs only with one final assembly node. In cases where a real assembly process consists of more than one final assembly station, then it is useful to split the assembly network into independent networks.
4. The minimum number of initial nodes i in the first tier t_0 is two.
5. Each node in the chain has at most one successor and must have at least two predecessors.
6. In a nonmodular ASC structure, the number of initial nodes i in the most upstream echelon is equal to the number of individual assembly parts or inputs.

Obtained unique topological structures are considered to be directed in tree graphs so that all edges point from left to right. Thus, they are convergent.

An example of the sets of structures for classes with numbers of initial nodes $i = 6$ and $i = 7$ can be found in Appendix 5.1, and ASC structures with numbers of initial nodes $i = 8$ can be found in Appendices 5.2 and 5.3.

5.2.2 *Complexity mitigation in assembly supply chain structures*

In Section 5.2.1, ASCs were modeled by unlabeled graphs. However, ASCs correspond with labeled graphs. Therefore, in order to mitigate the complexity of ASC structures, it is advisable to utilize the following steps:

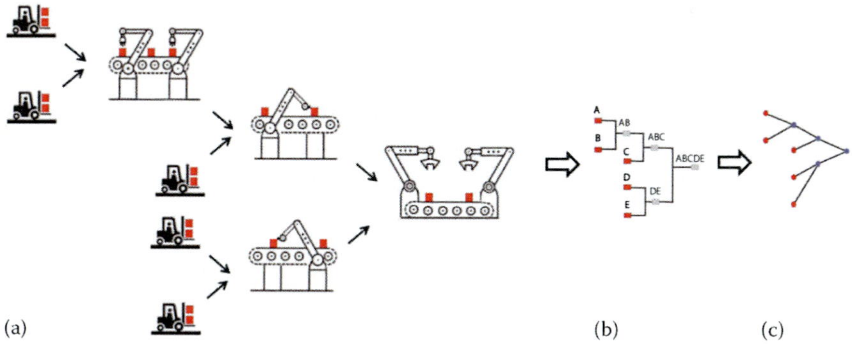

Figure 5.2 (a) Model of an ASC network; (b) transformation into a labeled graph; (c) transformation into ASC structure no. 10 in class #5.

1. Firstly, transform a realistic ASC system into a labeled graph and subsequently assign a corresponding ASC structure to this graph. A simple example is shown in Figure 5.2.
2. Subsequently, identify alternative ASC structures with the corresponding structure considered the original. Such alternative structures are available in all classes lower than the given ASC structure. As an example, we can use the original ASC structure from Figure 5.2c.

   Then, alternative ASC networks can be identified as depicted in Figure 5.3. Such alternative ASC networks are empirically less complex and at the same time keep the predefined assembly sequence unchanged. From a graph theory viewpoint, the obtained alternative graphs in the lower topological classes are then homeomorphic to the original labeled graph. From a practical point of view, such alternative ASC systems can be achieved based on an outsourcing method that often allows a company's clients to reach better, faster, and more sustainable results.
3. Finally, substitute intuitive methods for complexity comparison of benchmarked ASC structures by appropriate and effective indicators of topological complexity.

Coming back to Figure 5.3, four alternative ASC structures are taken from ASC topological classes #4, #3, and #2. Then, when considering, for example, the ASC structure taken from topological class #4 (namely, graph no. 5), it is assumed that assembly node DE with two external suppliers is substituted by one external supplier of the DE module. It is also evident that in alternative ASC models the predefined assembly sequences are unchanged. The arrangements of three other alternative ASC structures (graphs no. 4, no. 2, and no. 1) are analogical. In the next step, it is useful

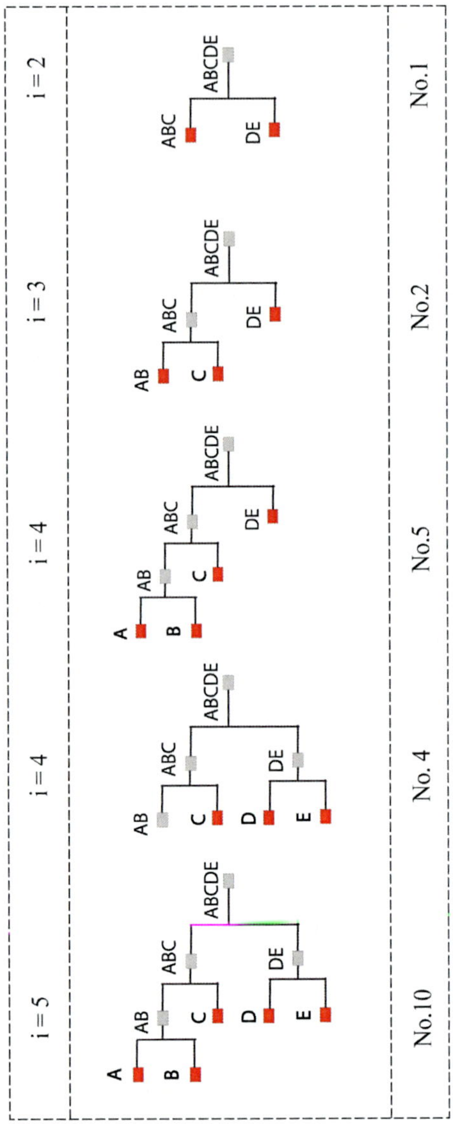

Figure 5.3 Original ASC structure where $i = 5$ and related alternative ASC networks.

to benchmark the complexity of the original ASC structure (graph no. 10) against the four alternative ASC networks. For this purpose, alternative complexity indicators are described in Section 5.3.

5.2.3 Approaches to the complexity of assembly supply chain structures

Metrics of structural complexity for specific or general networks can help us not only to design those systems but also better understand their topological properties. Layout design complexity metrics can be especially effective when comparing two or more ASC structures. Several studies in the literature [8–13] deal with complexity indicators for the structural complexity measurement of manufacturing systems. Since there are several types of ASC models, not all the complexity indicators are equally effective for different groups of networks. Respecting this fact, the following three complexity indicators are proposed.

5.2.3.1 Supply chain length

Nemeth and Foldesi [14] described the *supply chain length* (SCL) indicator and its extended definition. The SCL indicator takes, besides the number of nodes, the number of links weighted by their complexity into consideration. It is mainly focused on material flows. SCL is expressed by the equation:

$$\mathrm{SCL} = c_1.\sum_{i \in P} W_S.V_i + c_2.\sum_{(i,j) \in P} f\left(D_{i,j}\right).A_{i,j} \tag{5.1}$$

where:

- c_1 represents the technical and managerial level of vertices
- c_2 represents the technical and managerial level of edges
- w_S is the weight corresponding to the nature of the node
- P is the path from the origin to the destination
- V_i is the vertices (nodes) in the path
- $A_{i,j}$ is the arcs (edges) in the path
- $D_{i,j}$ is the distance in logistic terms (in this study it equals 1)
- $f(D_{i,j})$ is the weight determined by the distance in logistic terms

Subsequently, the following SCL indictor values for all ASC structures from the relevant topological classes have been obtained (Table 5.1).

5.2.3.2 Vertex degree index I_{vd}

According to Shannon's information theory, the entropy of information $H(\alpha)$ in describing a message of N system elements, distributed according

Table 5.1 SCL values of ASC structures for the relevant topological classes

	SCL values for individual ASC structures											
Class number	No. 1	No. 2	No. 3	No. 4	No. 5	No. 6	No. 7	No. 8	No. 9	No. 10	No. 11	No. 12
#5	11	13	13	13	15	15	15	15	15	17	17	17
#4	9	11	11	13	13	—	—	—	—	—	—	—
#3	7	9	—	—	—	—	—	—	—	—	—	—
#2	5	—	—	—	—	—	—	—	—	—	—	—

to some equivalence criterion α into k groups of $N_1, N_2,\ldots, N_k$ elements, is calculated by the formula:

$$H(\alpha) = -\sum_{i=1}^{k} p_i \log p_i = -\sum_{i=1}^{k} \frac{N_i}{N} \log_2 \frac{N_i}{N} \tag{5.2}$$

where p_i specifies the probability of the occurrence of the elements in the ith group.

Since it is of interest to characterize the entropy of information of a network, according to Shannon [15], it is possible to substitute symbols or system elements for the vertices.

In order to define the probability for a randomly chosen system element i, it is possible to formulate the general weight function as $p_i = w_i/\Sigma w_i$, assuming that $\Sigma p_i = 1$. Considering the system elements, the vertices, and supposing the weights assigned to each vertex are the corresponding vertex degrees, one easily distinguishes the null complexity of the totally disconnected graph from the high complexity of the complete graph. Then, the probability for a randomly chosen vertex i in the complete graph of V vertices to have a certain degree $deg(v)_i$ can be expressed by the formula:

$$p_i = \frac{deg(v)_i}{\sum_{i=1}^{V} deg(v)_i} \tag{5.3}$$

Shannon defines information as

$$I = H_{max} - H \tag{5.4}$$

where H_{max} is the maximum entropy that can exist in a system with the same number of elements.

Subsequently, the information entropy of a graph with a total weight W and vertex weights w_i can be expressed in the form of the equation:

$$H(W) = W\log_2 W - \sum_{i=1}^{V} w_i \log_2 w_i \tag{5.5}$$

Since the maximum entropy is when all $w_i = 1$, then:

$$H_{max} = W\log_2 W \tag{5.6}$$

By substituting $W = \sum_{i=1}^{V} deg(v)_i$ and $w_i = deg(v)_i$, the information content of the vertex degree distribution of a network, called the *vertex degree index* (I_{vd}), is derived by Bonchev and Buck [16] and is expressed as follows:

$$I_{vd} = -\sum_{i=1}^{V} deg(v)_i \log_2 deg(v)_i \tag{5.7}$$

Table 5.2 summarizes the value of the I_{vd} indicator for all ASC structures in classes #2 to #5.

5.2.3.3 *Axiomatic design-based complexity SDC*

The main definition of *axiomatic design* [17] states that any process can be seen in four main domains: process, functional, customer, and physical. The process consists of several steps and at the end results in structured relations between customer needs, functional requirements (FRs), and selected design parameters (DPs). These relations or dependencies between FRs and DPs within any design hierarchy can be expressed by the equation:

$$\text{FR} = [A]\ \text{DP} \tag{5.8}$$

where each element of the matrix [A] can be expressed as A = FR/DP. Equation 5.8 can be expressed as each FR on the product component depends on the specific DPs of the product specified by the customer, so that each dependency [A] can be understood as the existing relation of FR to DP. If, in the design matrix of any process, element [A] refers to "0," then FR does not relate to DP, and vice versa for "1," where there is a relation between DP and FR.

According to this approach, we indicate each initial node of the ASC model as FR (e.g., FR_1 to FR_{10} at C_{10}) and each subassembly vertex as DP (e.g., DP_1 to DP_3, depending on the specific ASC structure at C_{10}), shown in Figure 5.4a,b. This is because initial nodes practically represent company

Table 5.2 I_{vd} values of ASC structures for the relevant topological classes

Class number	I_{vd} Values for individual ASC structures											
	No. 1	No. 2	No. 3	No. 4	No. 5	No. 6	No. 7	No. 8	No. 9	No. 10	No. 11	No. 12
#5	11,61	13,61	12,75	12,75	14,75	14,75	14,75	14,26	14,26	16,26	16,26	16,26
#4	8,00	10,00	9,51	11,51	11,51	—	—	—	—	—	—	—
#3	4,75	6,75	—	—	—	—	—	—	—	—	—	—
#2	2,00	—	—	—	—	—	—	—	—	—	—	—

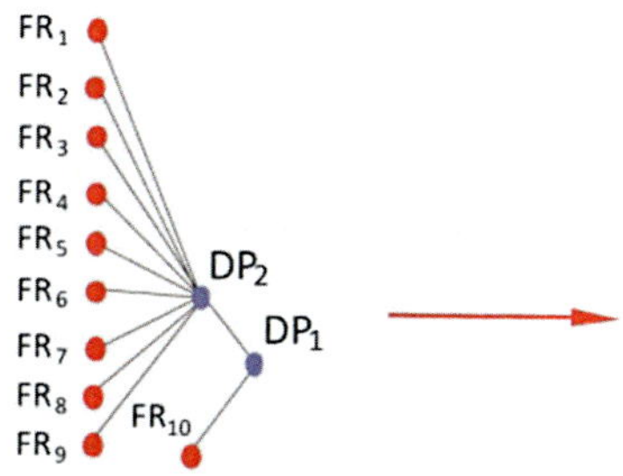

	DP_1	DP_2
FR_1	1	1
FR_2	1	1
FR_3	1	1
FR_4	1	1
FR_5	1	1
FR_6	1	1
FR_7	1	1
FR_8	1	1
FR_9	1	1
FR_{10}	1	0

(a)

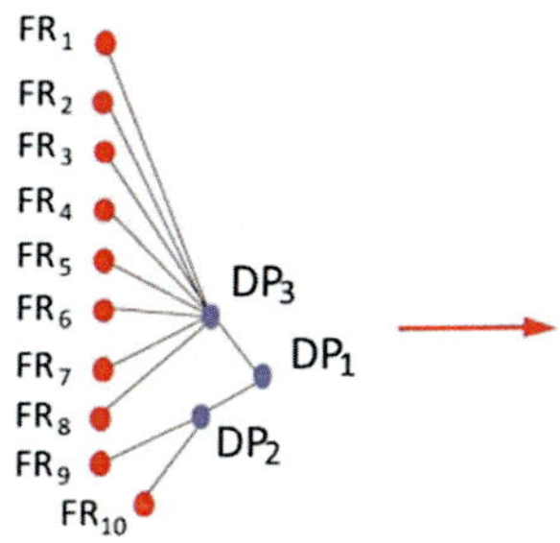

	DP_1	DP_2	DP_3
FR_1	1	0	1
FR_2	1	0	1
FR_3	1	0	1
FR_4	1	0	1
FR_5	1	0	1
FR_6	1	0	1
FR_7	1	0	1
FR_8	1	0	1
FR_9	1	1	0
FR_{10}	1	1	0

(b)

Figure 5.4 (a) ASC structure with 10 FRs and 2 DPs; (b) ASC structure with 10 FRs and 3 DPs, both transformed into a design matrix.

requirements on suppliers and specify the number of initial nodes in the ASC. Subsequently, the DPs are determined by these FRs as substructures. This way, the transformation of all repeated ASCs is possible and valuable.

Analogically, we can transform each ASC structure into an axiomatic design matrix (see the examples in Figure 5.4a for 10 FRs and two DPs and Figure 5.4b for 10 FRs and three DPs).

The presented design matrices have been transformed as coupled designs. For such matrices, it is characteristic that individual elements [*A*] are mostly nonzero, and thus the FRs cannot be satisfied independently.

5.2.4 *Comparison of approaches to structural complexity*

The aforementioned indicators can be assessed in view of their applicability as follows. As seen in Table 5.1, the values of the SCL indicator are identical for several structures in individual ASC classes #2 to #5. For example, class #5 with 12 individual ASC graphs can be divided into four SCL complexity levels. In the case of the I_{vd} indicator application, 12 graphs in the same ASC class #5 can be divided into six levels of structural complexity (see Table 5.2).

The two mentioned complexity indicators are therefore not the optimal measures of structural complexity of ASCs. On the other hand, the axiomatic design-based indicator SDC considers graphs links as interactions between nodes. Then, the complexity values of the SDC indicator for the same ASC structures in classes #2 to #5 differ for each of the graphs (Table 5.3).

Concluding the computational analysis using three structural complexity indicators, we may state that AD-based complexity best suits the given purpose.

A summary table of SCL, I_{vd}, and SDC complexity values for all ASC structures in classes from $i = 2$ to $i = 8$ can be found in Appendix 5.4.

5.3 *Modeling assembly supply chain structures for mass customization*

In Section 5.2.1, how to identify all possible ASC structures in order to select an optimal ASC structure was discussed. It was also assumed that ASCs correspond with labeled graphs, in which initial nodes represent stable assembly components. In terms of MC, these assembly components are commonly categorized into different types. Then, ASC structures are composed of at least one assembly module, with possible selection(s) from input components.

In order to outline the categorization of assembly components in MC supply chain systems, we consider exactly three types of initial components. They are as follows [18]:

1. *Stable components* are considered to be assembled to ensure the functionality of the module or final product.
2. *Voluntary optional components* are useful in some cases but not required. They can be selected by customers and are optional in any combination, including cases when only individual components are chosen. No component selection by customers is also an option.
3. *Compulsory optional components* differ from voluntary optional ones by the number of components that may be chosen from all of them; they are limited in selection. Thus, restrictions are determined by minimum, maximum, or exact requirements on a selection. The selection rules can be specified in a simple way by a combinatorial number

   $$\binom{k}{l}$$

   where l defines ways of picking component combinations from a set of all k, while $1 \leq l < k$.

Let us take the simple example of an ASC using structure number 10 from class #5 (Figure 5.2b), in which two types of initial components, stable and compulsory optional, will be considered (Figure 5.5).

Table 5.3 SDC values of ASC structures for the relevant topological classes

Class number	SDC values for individual ASC structures											
	No.1	No. 2	No. 3	No.4	No. 5	No. 6	No. 7	No. 8	No. 9	No. 10	No. 11	No. 12
#5	8,05	13,59	11,34	9,43	12,73	14,98	16,89	10,82	12,73	14,12	16,36	18,27
#4	5,55	8,84	6,93	8,32	10,23	—	—	—	—	—	—	—
#3	3,30	4,68	—	—	—	—	—	—	—	—	—	—
#2	1,38	—	—	—	—	—	—	—	—	—	—	—

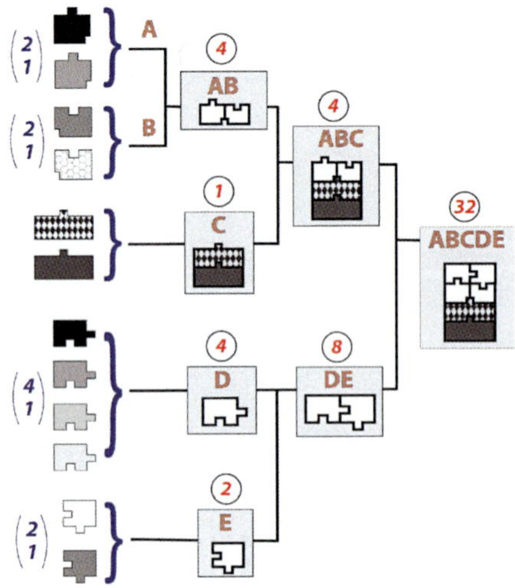

Figure 5.5 Labeled graph incorporating compulsory optional components.

Then, using simple combinatorial rules, we are able to obtain 32 design alternatives of product ABCDE. From the customers' perspective, such customized assembly offers 32 individual product configurations composed of 4 changeable assembly modules with 10 options for five initial parts (A–E). One can see that if an ASC consists of multiple initial component types and multiple modules, the enumeration of all possible product configurations is not easy. Accordingly, it would be useful to establish an effective framework for product variety quantification in terms of MC. Subsequently, it would be possible to handle so-called variety-induced complexity.

5.4 Conclusions

Complexity topology analysis of the ASC structures in Sections 5.2.3 and 5.2.4 revealed the potential tools to optimize the ASC structures to be used in an MC environment. Moreover, it has been found that modeling all possible ASCs is useful because any existing structure can be simplified using the approach presented in this chapter to obtain less complex ASC alternative(s). Secondly, the three complexity metrics—namely, SCL, I_{vd}, and SDC—used to capture the structural properties of all possible ASC networks have been benchmarked. One of them, the SDC indicator, best fits decision-making about the optimal supply chain as it also considers links between nodes and their interoperability.

Subsequently, in Section 5.3, it was shown that the proposed approach to model ASC networks can be effectively used in the MC manufacturing environment. Finally, a draft of the concept for product variety quantification has been outlined as a precondition for posterior solutions of product variety complexity mitigation.

References

1. Stump, B., and Badurdeen, F. (2012). Integrating lean and other strategies for mass customization manufacturing: A case study. *Journal of Intelligent Manufacturing*, 23(1), 109–124.
2. Brosch, M., Beckmann, G., and Krause, D. (2011). Approach to visualize the supply chain complexity induced by product variety. In *DS 68-5: Proceedings of the 18th International Conference on Engineering Design (ICED 11), Impacting Society through Engineering Design, Vol. 5: Design for X/Design to X*, Lyngby/Copenhagen, Denmark: Technical University of Denmark, 15–19.08.2011.
3. Blecker, T., Friedrich, G., Kaluza, B., Abdelkafi, N., and Kreutler, G. (2004). *Information and Management Systems for Product Customization*, Vol. 7. Berlin: Springer Science & Business Media.
4. Koren, Y., Heisel, U., Jovane, F., Moriwaki, T., Pritschow, G., Ulsoy, G., and Van Brussel, H. (1999). Reconfigurable manufacturing systems. *CIRP Annals: Manufacturing Technology*, 48(2), 527–540.
5. Wang, H. (2010). Product variety induced complexity and its impact on mixed-model assembly systems and supply chains. Doctoral dissertation, General Motors, University of Michigan, Ann Arbor, MI.
6. Zhu, X., Hu, S. J., Koren, Y., and Marin, S. P. (2008). Modeling of manufacturing complexity in mixed-model assembly lines. *Journal of Manufacturing Science and Engineering*, 130(5), 051013–10.
7. Modrak, V., and Marton, D. (2013). Development of metrics and a complexity scale for the topology of assembly supply chains. *Entropy*, 15(10), 4285–4299.
8. ElMaraghy, H., AlGeddawy, T., Samy, S. N., and Espinoza, V. (2014). A model for assessing the layout structural complexity of manufacturing systems. *Journal of Manufacturing Systems*, 33(1), 51–64.
9. Crippa, R., Bertacci, N., and Larghi, L. (2006). Representing and measuring flow complexity in the extended enterprise: The D4G approach. RIRL International Congress for Research in Logistics.
10. Frizelle, G., and Woodcock, E. (1995). Measuring complexity as an aid to developing operational strategy. *International Journal of Operations and Production Management*, 15(5), 26–39.
11. Deshmukh, A. V., Talavage, J. J., and Barash, M. M. (1998). Complexity in manufacturing systems, Part 1: Analysis of static complexity. *IIE Transactions*, 30(7), 645–655.
12. Espinoza, V. B. (2012). Vega structural complexity of manufacturing systems layout. MSc thesis, University of Windsor, ON.
13. Wang, H., Zhu, X., Hu, S. J., and Koren, Y. (2008). Complexity analysis of assembly supply chain configurations. In *ASME 2008 9th Biennial Conference on Engineering Systems Design and Analysis*, pp. 501–510. New York: American Society of Mechanical Engineers.
14. Németh, P., and Foldesi, P. (2009). Efficient control of logistic processes using multi-criteria performance measurement. *Acta Technica Jaurinensis*, 2, 353–360.

15. Shannon, C. E. (1948). A mathematical theory of communication. *Bell System Technical Journal*, 27, 379–423.
16. Bonchev, D., and Buck, G. A. (2005). Quantitative measures of network complexity. In D. Bonchev and D. H. Rouvray, Eds., *Complexity in Chemistry, Biology and Ecology*, Vol. 1, pp. 191–235. Berlin: Springer Science and Business Media.
17. Suh, N. P. (2005). Complexity in engineering. *CIRP Annals: Manufacturing Technology*, 54(2), 46–63.
18. Modrak, V., Marton, D., and Bednar, S. (2014). Modeling and determining product variety for mass customized manufacturing. *Procedia CIRP*, 23, 258–263.

Appendix 5.1

All possible ASC structures in class #6 and class #7.

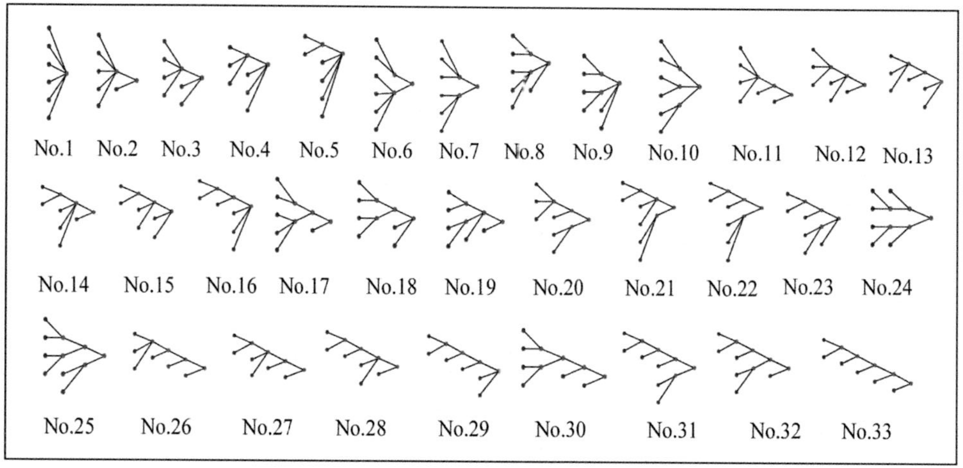

and

No.1
No.2
No.3
No.4
No.5
No.6
No.7
No.8
No.9
No.10
No.11
No.12
No.13
No.14
No.15
No.16
No.17
No.18
No.19
No.20
No.21
No.22
No.23
No.24
No.25
No.26
No.27
No.28
No.29
No.30
No.31
No.32
No.33
No.34
No.35
No.36
No.37
No.38
No.39
No.40
No.41
No.42
No.43
No.44
No.45
No.46
No.47
No.48
No.49
No.50
No.51
No.52
No.53
No.54
No.55
No.56
No.57
No.58
No.59
No.60
No.61
No.62
No.63
No.64
No.65
No.66
No.67
No.68
No.69
No.70
No.71
No.72
No.73
No.74
No.75
No.76
No.77
No.78
No.79
No.80
No.81
No.82
No.83
No.84
No.85
No.86
No.87
No.88
No.89
No.90

Appendix 5.2

All possible ASC structures in class #8 (part 1).

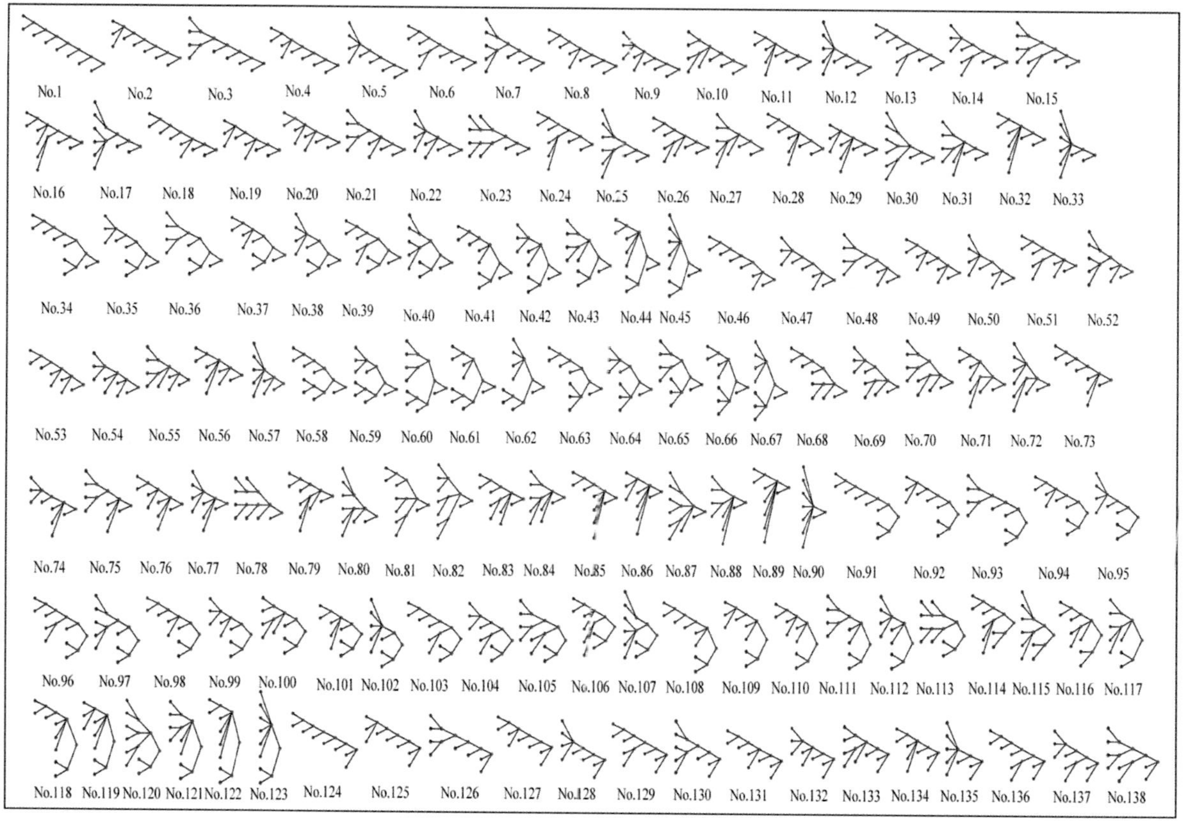

Appendix 5.3

All possible ASC structures in class #8 (part 2).

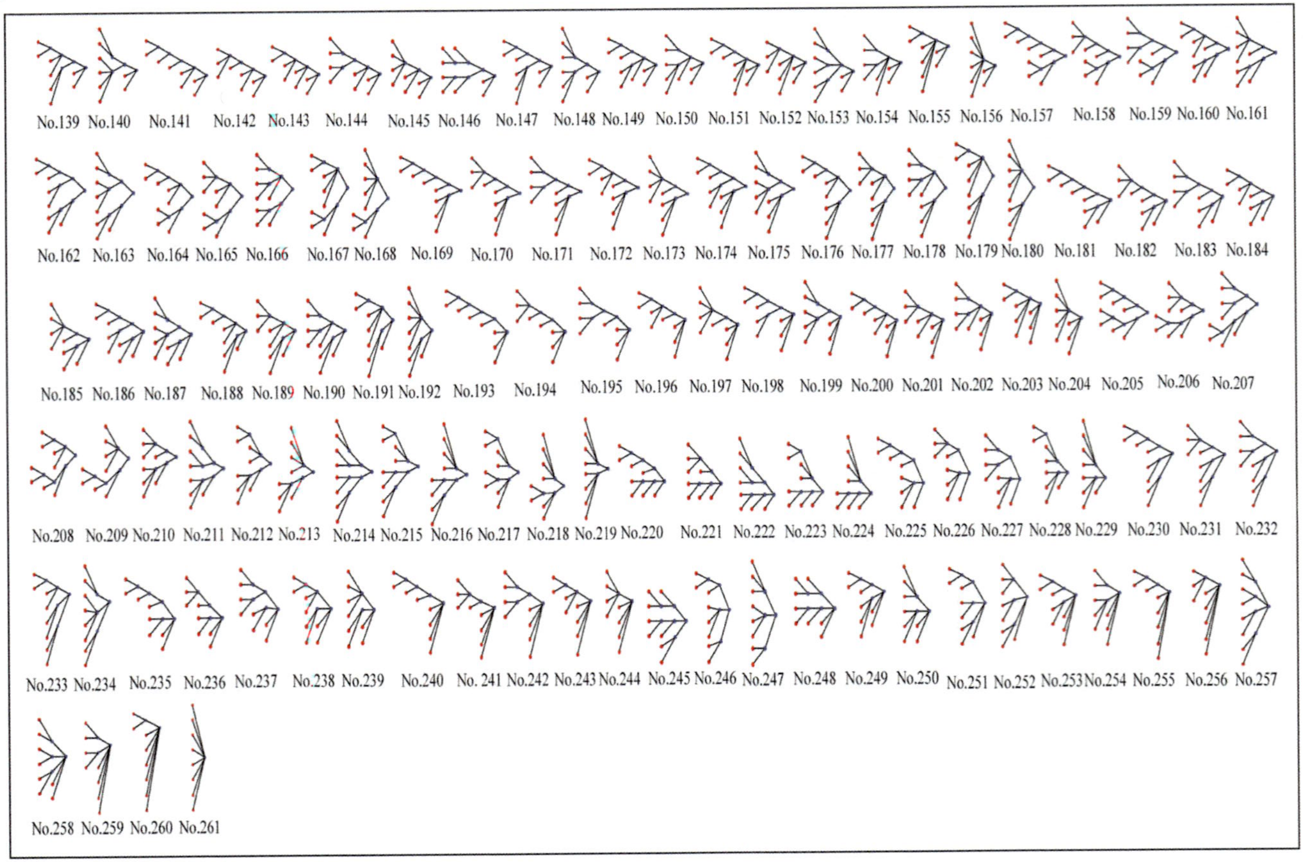

Appendix 5.4

Summary table of SCL, I_{vd}, and SDC complexity values for all ASC structures in classes $i = 2–8$.

i=2

deg(v)i	LSC	I_{vd}	SDC
2	5	2	1,39

i=3

deg(v)i	LSC	I_{vd}	SDC
3	7	4,75	3,3
3;2	9	6,75	4,68

i=4

deg(v)i	LSC	I_{vd}	SDC
4	9	8	5,55
4;2	11	10	8,84
3;3	11	9,51	6,93
3;3;2	13	11,51	8,32
3;3;2	13	11,51	10,23

i=5

deg(v)i	LSC	I_{vd}	SDC
5	11	11,61	8,05
5;2	13	13,61	13,59
4;3	13	12,75	11,34
4;3	13	12,75	9,43
4;3;2	15	14,74	12,73
4;3;2	15	14,74	14,98
4;3;2	15	14,74	16,89
3;3;3	15	14,26	10,8
3;3;3	15	14,26	12,73
3;3;3;2	17	16,26	14,12
3;3;3;2	17	16,26	16,36
3;3;3;2	17	16,26	18,27

i=6

deg(v)i	LSC	I_{vd}	SDC
6	13	15,51	10,75
6;2	15	17,51	17,70
5;3	15	16,36	12,14
5;3	15	16,36	16,30
5;3;2	17	18,36	17,68
5;3;2	17	18,36	24,34
5;3;2	17	18,36	20,18
4;4	15	16	14,05
4;4;2	17	18	17,34
4;4;2	17	18	22,09
4;3;3	17	17,51	15,43
4;3;3	17	17,51	13,52
4;3;3	17	17,51	19,59
4;3;3	17	17,51	17,68
4;3;3	17	17,51	15,43
4;3;3;2	19	19,51	23,48
4;3;3;2	19	19,51	21,57
4;3;3;2	19	19,51	20,98
4;3;3;2	19	19,51	19,07
4;3;3;2	19	19,51	18,73
4;3;3;2	19	19,51	27,64
4;3;3;2	19	19,51	25,73
4;3;3;2	19	19,51	23,48
3;3;3;3	19	19,02	14,91
3;3;3;3	19	19,02	19,07
3;3;3;3	19	19,02	16,82
3;3;3;3	19	19,02	20,98
3;3;3;3;2	21	21,02	20,11
3;3;3;3;2	21	21,02	20,45
3;3;3;3;2	21	21,02	27,12
3;3;3;3;2	21	21,02	22,36
3;3;3;3;2	21	21,02	24,87
3;3;3;3;2	21	21,02	29,03

i=7

deg(v)i	LSC	I_{vd}	SDC	deg(v)i	LSC	I_{vd}	SDC
7	15	19,65	13,62	4;3;3;3	21	22,26	23,85
7;2	17	21,65	24,37	4;3;3;3	21	22,26	21,94
6;3	17	20,26	21,67	4;3;3;3	21	22,26	21,60
6;3	17	20,26	15,01	4;3;3;3	21	22,26	19,69
6;3;2	19	22,26	32,42	4;3;3;3	21	22,26	19,69
6;3;2	19	22,26	25,76	4;3;3;3	21	22,26	17,78
6;3;2	19	22,26	23,05	4;3;3;3;2	23	24,26	41,26
5;4	17	19,61	19,17	4;3;3;3;2	23	24,26	39,35
5;4	17	19,61	16,92	4;3;3;3;2	23	24,26	37,10
5;4;2	19	21,61	29,92	4;3;3;3;2	23	24,26	37,10
5;4;2	19	21,61	27,67	4;3;3;3;2	23	24,26	35,19
5;4;2	19	21,61	22,46	4;3;3;3;2	23	24,26	34,60
5;3;3	19	21,12	27,21	4;3;3;3;2	23	24,26	32,69
5;3;3	19	21,12	23,05	4;3;3;3;2	23	24,26	34,60
5;3;3	19	21,12	20,55	4;3;3;3;2	23	24,26	32,69
5;3;3	19	21,12	18,30	4;3;3;3;2	23	24,26	32,35
5;3;3	19	21,12	16,39	4;3;3;3;2	23	24,26	30,44
5;3;3;2	21	23,12	37,96	4;3;3;3;2	23	24,26	28,53
5;3;3;2	21	23,12	33,81	4;3;3;3;2	23	24,26	31,90
5;3;3;2	21	23,12	31,30	4;3;3;3;2	23	24,26	29,99
5;3;3;2	21	23,12	29,05	4;3;3;3;2	23	24,26	27,74
5;3;3;2	21	23,12	27,14	4;3;3;3;2	23	24,26	27,74
5;3;3;2	21	23,12	28,60	4;3;3;3;2	23	24,26	25,83
5;3;3;2	21	23,12	24,44	4;3;3;3;2	23	24,26	27,14
5;3;3;2	21	23,12	23,85	4;3;3;3;2	23	24,26	25,23
4;4;3	19	20,75	24,96	4;3;3;3;2	23	24,26	27,14
4;4;3	19	20,75	22,46	4;3;3;3;2	23	24,26	25,23
4;4;3	19	20,75	20,55	3;3;3;3;3	23	23,77	31,90
4;4;3	19	20,75	20,21	3;3;3;3;3	23	23,77	29,99
4;4;3	19	20,75	18,30	3;3;3;3;3	23	23,77	27,74
4;4;3;2	21	22,75	35,71	3;3;3;3;3	23	23,77	25,23
4;4;3;2	21	22,75	33,21	3;3;3;3;3	23	23,77	23,33
4;4;3;2	21	22,75	31,30	3;3;3;3;3	23	23,77	22,99
4;4;3;2	21	22,75	30,96	3;3;3;3;3	23	23,77	21,08
4;4;3;2	21	22,75	29,05	3;3;3;3;3;2	25	25,77	42,65
4;4;3;2	21	22,75	26,35	3;3;3;3;3;2	25	25,77	40,74
4;4;3;2	21	22,75	25,76	3;3;3;3;3;2	25	25,77	38,49
4;4;3;2	21	22,75	23,85	3;3;3;3;3;2	25	25,77	35,99
4;3;3;3	21	22,26	30,51	3;3;3;3;3;2	25	25,77	34,08
4;3;3;3	21	22,26	28,60	3;3;3;3;3;2	25	25,77	33,74
4;3;3;3	21	22,26	26,35	3;3;3;3;3;2	25	25,77	33,28
4;3;3;3	21	22,26	26,35	3;3;3;3;3;2	25	25,77	31,37
4;3;3;3	21	22,26	24,44	3;3;3;3;3;2	25	25,77	29,12
4;3;3;3	21	22,26	23,85	3;3;3;3;3;2	25	25,77	28,53
4;3;3;3	21	22,26	21,94	3;3;3;3;3;2	25	25,77	26,62

i=8											
deg(v)i	*LSC*	I_{vd}	*SDC*	deg(v)i	*LSC*	I_{vd}	*SDC*	deg(v)i	*LSC*	I_{vd}	*SDC*
8	17	24	16,64	4;4;4	21	24	23,23	4;3;3;3;3;2	27	29,02	55,99
8;2	19	26	30,26	4;4;4;2	23	26	36,85	4;3;3;3;3;2	27	29,02	53,74
7;3	19	24,41	27,39	4;4;4;2	23	26	31,27	4;3;3;3;3;2	27	29,02	53,74
7;3	19	24,41	18,02	4;4;4;2	23	26	41,60	4;3;3;3;3;2	27	29,02	51,23
7;3;2	21	26,41	41,01	4;4;3;3	23	25,51	38,73	4;3;3;3;3;2	27	29,02	49,33
7;3;2	21	26,41	31,64	4;4;3;3	23	25,51	36,23	4;3;3;3;3;2	27	29,02	51,23
7;3;2	21	26,41	28,77	4;4;3;3	23	25,51	34,32	4;3;3;3;3;2	27	29,02	49,33
6;4	19	23,51	24,68	4;4;3;3	23	25,51	33,98	4;3;3;3;3;2	27	29,02	48,99
6;4	19	23,51	19,93	4;4;3;3	23	25,51	32,07	4;3;3;3;3;2	27	29,02	47,08
6;4;2	21	25,51	38,30	4;4;3;3	23	25,51	29,36	4;3;3;3;3;2	27	29,02	45,17
6;4;2	21	25,51	33,55	4;4;3;3	23	25,51	33,52	4;3;3;3;3;2	27	29,02	48,53
6;4;2	21	25,51	27,98	4;4;3;3	23	25,51	31,61	4;3;3;3;3;2	27	29,02	46,62
6;3;3	21	25,02	35,43	4;4;3;3	23	25,51	29,36	4;3;3;3;3;2	27	29,02	44,37
6;3;3	21	25,02	28,77	4;4;3;3	23	25,51	29,36	4;3;3;3;3;2	27	29,02	44,37
6;3;3	21	25,02	26,07	4;4;3;3	23	25,51	27,46	4;3;3;3;3;2	27	29,02	42,46
6;3;3	21	25,02	21,32	4;4;3;3	23	25,51	28,77	4;3;3;3;3;2	27	29,02	48,53
6;3;3	21	25,02	19,41	4;4;3;3	23	25,51	26,86	4;3;3;3;3;2	27	29,02	46,62
6;3;3;2	23	27,02	49,05	4;4;3;3	23	25,51	26,86	4;3;3;3;3;2	27	29,02	44,37
6;3;3;2	23	27,02	42,39	4;4;3;3	23	25,51	24,95	4;3;3;3;3;2	27	29,02	43,78
6;3;3;2	23	27,02	39,69	4;4;3;3	23	25,51	24,61	4;3;3;3;3;2	27	29,02	41,87
6;3;3;2	23	27,02	34,94	4;4;3;3	23	25,51	24,61	4;3;3;3;3;2	27	29,02	43,78
6;3;3;2	23	27,02	33,03	4;4;3;3	23	25,51	22,70	4;3;3;3;3;2	27	29,02	41,87
6;3;3;2	23	27,02	36,82	4;4;3;3;2	25	27,51	52,35	4;3;3;3;3;2	27	29,02	41,87
6;3;3;2	23	27,02	30,16	4;4;3;3;2	25	27,51	51,83	4;3;3;3;3;2	27	29,02	39,96
6;3;3;2	23	27,02	29,36	4;4;3;3;2	25	27,51	49,85	4;3;3;3;3;2	27	29,02	39,62
5;5	19	23,22	22,18	4;4;3;3;2	25	27,51	47,94	4;3;3;3;3;2	27	29,02	37,71
5;5;2	21	25,22	27,73	4;4;3;3;2	25	27,51	47,60	4;3;3;3;3;2	27	29,02	45,66
5;5;2	21	25,22	35,80	4;4;3;3;2	25	27,51	45,69	4;3;3;3;3;2	27	29,02	43,75
5;4;3	21	24,36	32,93	4;4;3;3;2	25	27,51	42,99	4;3;3;3;3;2	27	29,02	41,50
5;4;3	21	24,36	30,68	4;4;3;3;2	25	27,51	47,15	4;3;3;3;3;2	27	29,02	41,50
5;4;3	21	24,36	30,23	4;4;3;3;2	25	27,51	45,24	4;3;3;3;3;2	27	29,02	39,59
5;4;3	21	24,36	26,07	4;4;3;3;2	25	27,51	42,99	4;3;3;3;3;2	27	29,02	39,00
5;4;3	21	24,36	25,48	4;4;3;3;2	25	27,51	42,99	4;3;3;3;3;2	27	29,02	37,09
5;4;3	21	24,36	23,57	4;4;3;3;2	25	27,51	41,08	4;3;3;3;3;2	27	29,02	39,00
5;4;3	21	24,36	25,48	4;4;3;3;2	25	27,51	42,39	4;3;3;3;3;2	27	29,02	37,09
5;4;3	21	24,36	23,57	4;4;3;3;2	25	27,51	40,48	4;3;3;3;3;2	27	29,02	36,75
5;4;3	21	24,36	21,32	4;4;3;3;2	25	27,51	40,48	4;3;3;3;3;2	27	29,02	34,84
5;4;3;2	23	26,36	46,55	4;4;3;3;2	25	27,51	38,57	4;3;3;3;3;2	27	29,02	32,93
5;4;3;2	23	26,36	44,30	4;4;3;3;2	25	27,51	38,23	4;3;3;3;3;2	27	29,02	38,21
5;4;3;2	23	26,36	43,85	4;4;3;3;2	25	27,51	36,33	4;3;3;3;3;2	27	29,02	36,30
5;4;3;2	23	26,36	39,69	4;4;3;3;2	25	27,51	40,12	4;3;3;3;3;2	27	29,02	34,05
5;4;3;2	23	26,36	39,10	4;4;3;3;2	25	27,51	37,61	4;3;3;3;3;2	27	29,02	34,05
5;4;3;2	23	26,36	37,19	4;4;3;3;2	25	27,51	35,70	4;3;3;3;3;2	27	29,02	32,14
5;4;3;2	23	26,36	39,10	4;4;3;3;2	25	27,51	35,36	4;3;3;3;3;2	27	29,02	38,21
5;4;3;2	23	26,36	37,19	4;4;3;3;2	25	27,51	33,45	4;3;3;3;3;2	27	29,02	36,30
5;4;3;2	23	26,36	34,94	4;4;3;3;2	25	27,51	32,66	4;3;3;3;3;2	27	29,02	34,05
5;4;3;2	23	26,36	34,32	4;4;3;3;2	25	27,51	36,82	4;3;3;3;3;2	27	29,02	35,70
5;4;3;2	23	26,36	32,07	4;4;3;3;2	25	27,51	34,91	4;3;3;3;3;2	27	29,02	33,79
5;4;3;2	23	26,36	33,52	4;4;3;3;2	25	27,51	32,66	4;3;3;3;3;2	27	29,02	33,79
5;4;3;2	23	26,36	29,36	4;4;3;3;2	25	27,51	32,66	4;3;3;3;3;2	27	29,02	31,88
5;4;3;2	23	26,36	31,02	4;4;3;3;2	25	27,51	30,75	3;3;3;3;3;3	27	28,53	45,66
5;4;3;2	23	26,36	29,11	4;4;3;3;2	25	27,51	34,32	3;3;3;3;3;3	27	28,53	43,75
5;3;3;3	23	25,87	40,98	4;4;3;3;2	25	27,51	32,41	3;3;3;3;3;3	27	28,53	41,50
5;3;3;3	23	25,87	36,82	4;4;3;3;2	25	27,51	30,50	3;3;3;3;3;3	27	28,53	39,00
5;3;3;3	23	25,87	34,32	4;3;3;3;3	25	27,02	44,27	3;3;3;3;3;3	27	28,53	37,09
5;3;3;3	23	25,87	32,07	4;3;3;3;3	25	27,02	42,36	3;3;3;3;3;3	27	28,53	36,75
5;3;3;3	23	25,87	30,16	4;3;3;3;3	25	27,02	40,12	3;3;3;3;3;3	27	28,53	36,30
5;3;3;3	23	25,87	31,61	4;3;3;3;3	25	27,02	40,12	3;3;3;3;3;3	27	28,53	34,39
5;3;3;3	23	25,87	27,46	4;3;3;3;3	25	27,02	38,21	3;3;3;3;3;3	27	28,53	32,14
5;3;3;3	23	25,87	26,86	4;3;3;3;3	25	27,02	37,61	3;3;3;3;3;3	27	28,53	31,54
5;3;3;3	23	25,87	24,95	4;3;3;3;3	25	27,02	35,70	3;3;3;3;3;3	27	28,53	29,64
5;3;3;3	23	25,87	26,86	4;3;3;3;3	25	27,02	37,61	3;3;3;3;3;3	27	28,53	29,64
5;3;3;3	23	25,87	24,95	4;3;3;3;3	25	27,02	35,70	3;3;3;3;3;3	27	28,53	27,73
5;3;3;3	23	25,87	22,70	4;3;3;3;3	25	27,02	35,36	3;3;3;3;3;3	27	28,53	27,39
5;3;3;3	23	25,87	20,79	4;3;3;3;3	25	27,02	33,45	3;3;3;3;3;3;2	29	30,53	59,28
5;3;3;3;2	25	27,87	54,60	4;3;3;3;3	25	27,02	31,54	3;3;3;3;3;3;2	29	30,53	57,37
5;3;3;3;2	25	27,87	50,44	4;3;3;3;3	25	27,02	34,91	3;3;3;3;3;3;2	29	30,53	55,12
5;3;3;3;2	25	27,87	47,94	4;3;3;3;3	25	27,02	33,00	3;3;3;3;3;3;2	29	30,53	52,62
5;3;3;3;2	25	27,87	45,69	4;3;3;3;3	25	27,02	30,75	3;3;3;3;3;3;2	29	30,53	50,71
5;3;3;3;2	25	27,87	43,78	4;3;3;3;3	25	27,02	30,75	3;3;3;3;3;3;2	29	30,53	50,37
5;3;3;3;2	25	27,87	45,24	4;3;3;3;3	25	27,02	28,84	3;3;3;3;3;3;2	29	30,53	49,92
5;3;3;3;2	25	27,87	41,08	4;3;3;3;3	25	27,02	34,91	3;3;3;3;3;3;2	29	30,53	48,01
5;3;3;3;2	25	27,87	40,48	4;3;3;3;3	25	27,02	33,00	3;3;3;3;3;3;2	29	30,53	45,76
5;3;3;3;2	25	27,87	40,48	4;3;3;3;3	25	27,02	30,75	3;3;3;3;3;3;2	29	30,53	45,17
5;3;3;3;2	25	27,87	38,57	4;3;3;3;3	25	27,02	30,16	3;3;3;3;3;3;2	29	30,53	43,26
5;3;3;3;2	25	27,87	36,33	4;3;3;3;3	25	27,02	28,25	3;3;3;3;3;3;2	29	30,53	47,05
5;3;3;3;2	25	27,87	34,42	4;3;3;3;3	25	27,02	30,16	3;3;3;3;3;3;2	29	30,53	45,14
5;3;3;3;2	25	27,87	42,36	4;3;3;3;3	25	27,02	28,25	3;3;3;3;3;3;2	29	30,53	42,89
5;3;3;3;2	25	27,87	38,21	4;3;3;3;3	25	27,02	28,25	3;3;3;3;3;3;2	29	30,53	40,39
5;3;3;3;2	25	27,87	35,70	4;3;3;3;3	25	27,02	26,34	3;3;3;3;3;3;2	29	30,53	38,48
5;3;3;3;2	25	27,87	33,45	4;3;3;3;3	25	27,02	28,25	3;3;3;3;3;3;2	29	30,53	38,14
5;3;3;3;2	25	27,87	31,54	4;3;3;3;3	25	27,02	26,34	3;3;3;3;3;3;2	29	30,53	39,59
5;3;3;3;2	25	27,87	34,91	4;3;3;3;3	25	27,02	26,00	3;3;3;3;3;3;2	29	30,53	37,68
5;3;3;3;2	25	27,87	30,75	4;3;3;3;3	25	27,02	26,00	3;3;3;3;3;3;2	29	30,53	35,43
5;3;3;3;2	25	27,87	32,41	4;3;3;3;3	25	27,02	24,09	3;3;3;3;3;3;2	29	30,53	37,09
5;3;3;3;2	25	27,87	30,50	4;3;3;3;3	25	27,02	22,18	3;3;3;3;3;3;2	29	30,53	35,18
4;4;4	21	24	27,98	4;3;3;3;3;2	27	29,02	57,90	3;3;3;3;3;3;2	29	30,53	33,27

chapter six

Variety-induced complexity metrics

Vladimir Modrak and Slavomir Bednar

Contents

ABSTRACT
It is frequently discussed that product variety impacts manufacturing complexity, and in this context the assessment of variety-induced complexity is becoming a topical problem. Then, there is an implicit need for configuration complexity management methods to support managers in taking proper decisions. The main scope of this chapter is a description of the combinatorial-based method to quantify product configurations and

variations arising in mass customized manufacturing. In this order, two scenarios of structuring and quantifying product components will be presented. The first one considers two basic types of input assembly components—namely, stable and optional. The second scenario is enlarged with an additional component type, which is the compulsory optional component. Finally, the numbers of all possible product configurations and newly proposed axiomatic design (AD)-based measures are benchmarked on a model of the mass customized assembly (MCA) process. As a result, both measures can be effectively used to assist product managers and marketing advisors to independently assess alternative product structures.

6.1 Complexity as an important factor influencing mass customization strategy

Product configuration complexity problems are frequently discussed in connection with mass customization (MC). Moreover, product complexity management is considered an important topic among company managers and academics alike. At the same time, it is not easy to precisely define the complexity of product configuration options. Tihonen et al. [1] argue that the complexity of product configuration options is directly related to the degree of modularity of the product. Probably, any definition of the product configuration complexity would be necessarily beholden to different types of products, as there are many definitions of the term *product configuration* [2–4]. However, the ability to measure the complexity of product configuration using a reliable variety-based metric would allow benchmarking of the concurrent or alternative product variety platforms. It is possible to identify several pertinent facets of complexity in this domain. Calinescu et al. [5] provided a comprehensive view on the various aspects of manufacturing complexity, including product structure complexity. According to Jiao et al. [6], a product structure is defined in terms of module types, while product variants derived from this product structure share the same module types and take on different instances of every module type. Liu et al. [7] defined product structure by tree levels—for example, systems, subsystems, and modules—in which components can be added, modified, or deleted at each level. Fredendall and Gabriel [8] pointed out that there are four building elements of the product structure complexity that seem to contribute especially to product variety complexity. These are (1) the number of assembled components on entry to an MCA system, (2) the number of manufactured items, (3) the number of levels in a product assembly structure, and (4) the degree of part commonality. It is evident that the number of assembled components and the number of manufactured products definitely increases the complexity of scheduling and even material control. But the number of components on entry to the assembly system and the number of levels in the product structure independently contribute to the level of product variety complexity.

The complexity of any system is affected mainly by three variables: namely, the state of the system elements, their number, and the relationships among them. Several definitions of manufacturing complexity have been provided so far but the very first definition is associated with Shannon's information theory [9], related to the amount of information (in bits) in the uncertainty of the information system. From this approach, it is evident that the fewer processes, machines, and/or product configurations there are, the lower the overall complexity of the system. Zhu et al. [10] proposed a measure of complexity based on the choices that the operator has to make at the station level. According to Desmukh et al. [11], product design modifications can have a significant impact on the static complexity of the manufacturing system.

Suh [12] defined complexity in relation to product design through the achievement of functional and design requirements. Kim et al. [13] introduced a number of metrics for complexity on the basis of system components, elements, and their relations. These measures cover the majority of system elements but cannot be extended to other manufacturing domains, except for cell production. Frizelle and Woodcock [14] defined two original types of complexity, static and dynamic, currently corresponding with structural and operational complexity. Later, Frizelle and Efstathiou [15] presented the former as *good complexity* and the latter *bad complexity*. Their metrics have been further applied and even developed in terms of MC by other authors [16,17].

Kampker et al. [18] categorized the term *product variety* into two types: internal and external. At the same time, they emphasized that "product architecture and technology determine the ratio between the variety externally offered to the market and internally produced." According to Grussenmeyer and Blecker [19], there is a need for a novel complexity management method to support managers in taking proper new product development decisions for the selection of suitable product variety platforms.

Our efforts in this chapter will focus on the measurement and assessment of the product configuration options from a complexity viewpoint. At the beginning, combinatorial-based methods to quantify product configurations and variations representing product variety extent will be outlined.

Two scenarios that differ in the presence of initial component types will be assumed. The first one (Scenario #1) considers two basic types of input components for assembly operation—namely, stable (S) and voluntary optional (VO) components. The second (Scenario #2) is enlarged with an additional component type—namely, compulsory optional (CO) components.

6.2 Quantification of product configurations and variations for Scenario #1

6.2.1 The conceptual model of product configurations

A quantification of product configurations for the two component types assumes the following number of input components: an unlimited number of S components *i*, starting with $i = 1$, and an unlimited number of VO components *j*, starting with $j = 0$. In order to outline the model of product configurations, the following simple example will be used under the assumption that the input components will be assembled into a submodule, module, or product, respectively. Let us have two S components ($i = 2$) and two VO components ($j = 2$). This customizable assembly operation can be categorized by the notation CL_iSCL_j, where CL_i refers to the *class* of S components with the number of S components *i*, and SCL_j refers to the *subclass* of voluntary components with the number of voluntary components *j*. Thus, the example belongs to class and subclass CL_2SCL_2. Then, four product configurations may occur (Figure 6.1).

If the number of S components is $i \geq 2$, and if zero VO are selected, we identify one product configuration, as shown in Figure 6.1.

6.2.2 Conceptual model of product variations

In the case of product configuration modeling, the permutations of input components have been omitted. But if we consider that permutations among components are relevant, then it is possible to generate for each class and subclass of product configuration a related number of so-called product variations. One may determine a mathematical expression for the quantification of product variations. These variations are pertinent if any product configuration consists of at least one VO component and the number of S components $i \geq 1$.

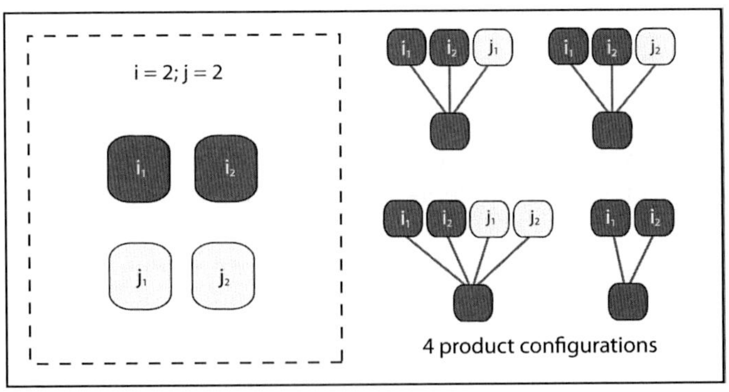

Figure 6.1 MCA illustration for CL_2SCL_2.

However, product variations may have more or less only theoretical importance; therefore, their relevance will be proved in Section 6.2.4.

In order to introduce product variations and their quantification, let us use a case with three S input components ($i = 3$) and two VO input components ($j = 2$). Then, four product configurations can be obtained (Figure 6.2a).

Each of the four product configurations in Figure 6.2a may include a number of exactly related product variations based on the permutation rules. For example, a product configuration where $i = 3$ and $j = 2$ generates nine product variations, as depicted in Figure 6.2b. In summary, the whole MCA node CL_3SCL_2 offers 16 product variations.

6.2.3 Combinatorial formulas for determining product configurations

Following the principles of combinatorics as described, we have established formulas for the determination of product configurations. Two types of enumerations have been identified. The first of them is for the case CL_1SCL_j and the second one is for the case $CL_{2-\infty}SCL_j$. For each of these types, there are two different ways to determine the number of product configurations (NPC). The first of them brings only the total NPC, but the second way of looking at it is through the structure of product configurations. For example, we know that the case CL_3SCL_2 generates four product configurations, but we do not know the structure and distribution of these configurations. The second way of calculation allows us to also know the structure of these configurations.

Let us further use both methods of enumeration for the case $CL_1SCL_{0-\infty}$, as follows:

1. When the structure of the product configurations is unimportant, the elementary formula for quantifying the product configurations using the first way will be

$$\sum Conf_{CL_1SCL_j} = \left(2^j\right) - 1 \tag{6.1}$$

   In this formula, a configuration with a single S component is not considered an assembly operation and therefore this configuration is subtracted.
2. When the structure of the product configurations is important, then, for case $CL_1SCL_{0-\infty}$, the formula is

$$\sum Conf_{CL_1SCL_j} = \sum_{j=1}^{n_j} \left(\frac{j!}{n_j!(j-n_j)!} \right) \tag{6.2}$$

   where n_j is a set of integers limited to a range 1 to j.

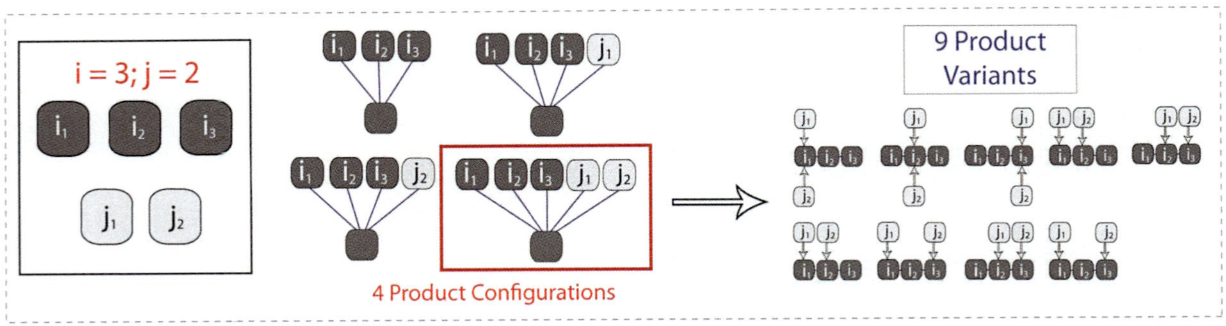

Figure 6.2 (a) Four possible product configurations for product class CL_3SCL_2; (b) nine possible product variations for the selected product configurations.

The second type of enumeration for the case $CL_{2-\infty}SCL_{0-\infty}$ is as follows:

1. When the structure of product configurations is unimportant, the formula is

$$\sum Conf_{CL_{2-\infty}SCL_j} = \left(2^j\right) \tag{6.3}$$

2. When the structure of product configurations is important, the formula is

$$\sum Conf_{CL_{2-\infty}SCL_j} = \sum_{j=0}^{n_j}\left(\frac{j!}{n_j!(j-n_j)!}\right) \tag{6.4}$$

where n_j is a set of integers limited to a range 0 to j.

An example of how Formula 6.4 can be applied is shown in simple case composition by using class and subclass CL_4SCL_3:

$$\sum Conf_{CL_4SCL_3} = \left(\frac{3!}{0!(3-0)!}\right)+\left(\frac{3!}{1!(3-1)!}\right)+\left(\frac{3!}{2!(3-2)!}\right)+\left(\frac{3!}{3!(3-3)!}\right)$$
$$=1+3+3+1=8$$

This calculation offers a view on how j component combinations are distributed. In our case, it shows that we obtain one component combination consisting only of S components, three combinations with one optional component, three combinations with two optional components, and one component combination consisting of all optional components, as graphically shown in Figure 6.3.

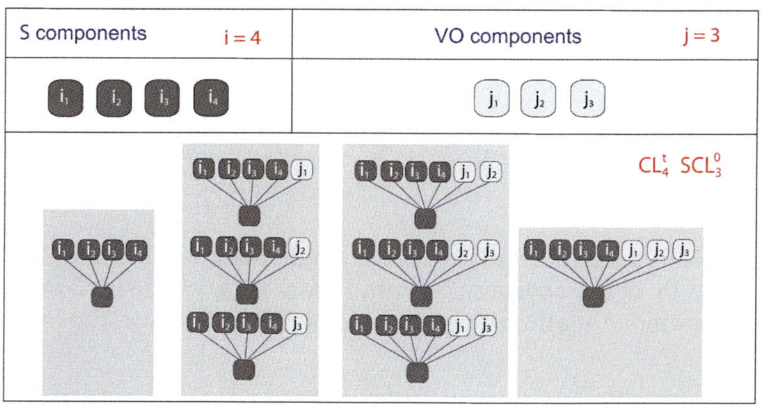

Figure 6.3 Distribution of product configurations in calculation for the product class CL_4SCL_3.

In cases where a distribution of product configurations is unimportant, we can apply Formula 6.3, by which we obtain the same result:

$$\sum Conf_{CL_4SCL_3^0} = 2^3 = 8$$

6.2.4 *Combinatorial approach to determining product variations*

As the title of this subsection indicates, after determining the product configurations, it is possible to identify a related number of product variations for each product configuration. Doing so, two basic approaches can be applied.

The first way to enumerate a number of product variations (ΣVar) for any class CL_i and subclass SCL_j is by taking any single product configuration. Then, the formula for the enumeration of product variations is as follows:

$$\sum Var_{CL_iSCL_j} = \left(\frac{j!i^{n_j}}{n_{j!}(j-n_j)!} \right) \tag{6.5}$$

where n_j is a set of integers limited to a range 0 to j.

The second way is pertinent when we need to quantify a sum of product variations for all possible product configurations of the given class CL_i and subclass SCL_j. Then, the formula for the quantification of product variations in class CL_1 is as follows:

$$\sum Var_{CL_iSCL_j} = \sum_{j=1}^{n_j} \left(\frac{j!i^{n_j}}{n_{j!}(j-n_j)!} \right) \tag{6.6}$$

An expression for the quantification of product variations in class $CL_{2-\infty}$ is

$$\sum Var_{CL_iSCL_j} = \sum_{j=0}^{n_j} \left(\frac{j!i^{n_j}}{n_{j!}(j-n_j)!} \right) \tag{6.7}$$

The validity of Formula 6.5 for a single product configuration can be proved by using an example from Figure 6.2b, where $i = 3$ and $j = 2$, then the number of all product variations is calculated in the following way:

$$\sum Var_{CL_3SCL_2} = \left(\frac{2!3^2}{2(2-2)!} \right) = 9$$

Subsequently, the number of product variations for all relevant product configurations in Figure 6.2a can be obtained through Formula 6.7:

$$\sum Var_{CL_3SCL_2} = \left(\frac{2!3^0}{0!(2-0)!}\right) + \left(\frac{2!3^1}{1!(2-1)!}\right) + \left(\frac{2!3^2}{2!(2-2)!}\right) = 16$$

Summarizing the results of the selected individual node of assembly operation with $i = 3$ and $j = 2$, four product configurations and 16 product variations were identified.

Accordingly, for any composition of entry components, the product configurations and the related number of product variations can be quantified.

From the results of the numerical experiments offered in Table 6.1, it is shown that product variations grow unrealistically while product configurations grow in a way that is closer to real situations in terms of MC.

For example, if $i = 5$ and $j = 6$, then 48 product configurations and 46,656 product variations can be composed from this initial component set. However, empirical experiences tell us that the number of demanded product configurations is usually lower than the number of all possible product configurations. Therefore, a variety extent expressed through the number of all possible product variations might be more or less considered only as a theoretical construct. Further, we provide extended summary tables for both product configurations and product variations in the classes and subclasses CL_1SCL_{1-36} and CL_2SCL_{0-35} (Tables 6.2 and 6.3).

From Tables 6.2 and 6.3, one can see that in the case of a single S component, the numbers of product configurations and product variations are identical. If we have one generic product—for example, a specific car model with 10 supplementary equipment options—then the maximum number of product configurations and variations is 1023. But in cases when the number of S components $i \geq 2$, then product variations have strong exponential growth, while product configurations grow more steadily.

6.3 *Quantification of product configurations for Scenario #2*

In contrast with Scenario #1, where both product configurations and product variations were analyzed, this scenario will treat only product configurations.

In Scenario #2, a quantification of product configurations is enriched with a CO component type. Then, a notation for a subclass of voluntary components SCL_j^k will include new information through index k, which determines the number of CO components in the assembly component

Table 6.1 Related values of product configurations and variations for classes CL_{2-5}

CL_i	SCL_j	I	j	$\Sigma Conf$	ΣVar
CL_2	SCL_0	2	0	1	1
	SCL_1	2	1	2	3
	SCL_2	2	2	4	9
	SCL_3	2	3	8	27
	SCL_4	2	4	16	81
	SCL_5	2	5	32	243
	SCL_6	2	6	48	729
CL_3	SCL_0	3	0	1	1
	SCL_1	3	1	2	4
	SCL_2	3	2	4	16
	SCL_3	3	3	8	64
	SCL_4	3	4	16	256
	SCL_5	3	5	32	1,024
	SCL_6	3	6	48	4,096
CL_4	SCL_0	4	0	1	1
	SCL_1	4	1	2	5
	SCL_2	4	2	4	25
	SCL_3	4	3	8	125
	SCL_4	4	4	16	625
	SCL_5	4	5	32	3,125
	SCL_6	4	6	48	15,625
CL_5	SCL_0	5	0	1	1
	SCL_1	5	1	2	6
	SCL_2	5	2	4	36
	SCL_3	5	3	8	216
	SCL_4	5	4	16	1,296
	SCL_5	5	5	32	7,776
	SCL_6	5	6	48	46,656

combination. Such component types frequently occur in many customizable assembly operations. Requirements on the selection of components from CO components are expressed through a variable l. Here, it is assumed that l as an integer is limited by the rule $1 \leq l < k$.

Then, it is necessary to specify the rules for the selection of CO components. They are as follows:

- Rule A: Individual selectivity rule

  We may define the exact number l of components to be chosen from all k of CO components.

Table 6.2 Summary table with related values of product configurations ($\Sigma Conf$) and variations (ΣVar) for product class CL_1

Class	Components (S)	Subclass	Components (VO)	$\Sigma Conf$	ΣVar
CL_1	1	SCL_1	1	1	1
		SCL_2	2	3	3
		SCL_3	3	7	7
		SCL_4	4	15	15
		SCL_5	5	31	31
		SCL_6	6	63	63
		SCL_7	7	127	127
		SCL_8	8	255	255
		SCL_9	9	511	511
		SCL_{10}	10	1,023	1,023
		SCL_{11}	11	2,047	2,047
		SCL_{12}	12	4,095	4,095
		SCL_{13}	13	8,191	8,191
		SCL_{14}	14	16,383	16,383
		SCL_{15}	15	32,767	32,767
		SCL_{16}	16	65,535	65,535
		SCL_{17}	17	131,071	131,071
		SCL_{18}	18	262,143	262,143
		SCL_{19}	19	524,287	524,287
		SCL_{20}	20	1,048,575	1,048,575
		SCL_{21}	21	2,097,151	2,097,151
		SCL_{22}	22	4,194,303	4,194,303
		SCL_{23}	23	8,388,607	8,388,607
		SCL_{24}	24	16,777,215	16,777,215
		SCL_{25}	25	33,554,431	33,554,431
		SCL_{26}	26	67,108,863	67,108,863
		SCL_{27}	27	134,217,727	134,217,727
		SCL_{28}	28	268,435,455	268,435,455
		SCL_{29}	29	536,870,911	536,870,911
		SCL_{30}	30	1,073,741,823	1,073,741,823
		SCL_{31}	31	2,147,483,647	2,147,483,647
		SCL_{32}	32	4,294,967,295	4,294,967,295
		SCL_{33}	33	8,589,934,591	8,589,934,591
		SCL_{34}	34	17,179,869,183	17,179,869,183
		SCL_{35}	35	34,359,738,367	34,359,738,367
		SCL_{36}	36	68,719,476,735	68,719,476,735

Table 6.3 Summary table with related values of product configurations ($\Sigma Conf$) and variations (ΣVar) for product class $CL_{2-\infty}$

Class	Subclass	$\Sigma Conf$	ΣVar
CL_2	SCL_1	2	3
	SCL_2	4	9
	SCL_3	8	27
	SCL_4	16	81
	SCL_5	32	243
	SCL_6	64	729
	SCL_7	128	2,187
	SCL_8	256	6,561
	SCL_9	512	19,683
	SCL_{10}	1,024	59,049
	SCL_{11}	2,048	177,147
	SCL_{12}	4,096	531,441
	SCL_{13}	8,192	1,594,323
	SCL_{14}	16,384	4,782,969
	SCL_{15}	32,768	14,348,907
	SCL_{16}	65,536	43,046,721
	SCL_{17}	131,072	129,140,163
	SCL_{18}	262,144	387,420,489
	SCL_{19}	524,288	1,162,261,467
	SCL_{20}	1,048,576	3,486,784,401
	SCL_{21}	2,097,152	10,460,353,203
	SCL_{22}	4,194,304	31,381,059,609
	SCL_{23}	8,388,608	94,143,178,827
	SCL_{24}	16,777,216	282,429,536,481
	SCL_{25}	33,554,432	847,288,609,443
	SCL_{26}	67,108,864	2,541,865,828,329
	SCL_{27}	134,217,728	7,625,597,484,987
	SCL_{28}	268,435,456	22,876,792,454,961
	SCL_{29}	536,870,912	68,630,377,364,883
	SCL_{30}	1,073,741,824	205,891,132,094,649
	SCL_{31}	2,147,483,648	617,673,396,283,947
	SCL_{32}	4,294,967,296	1,853,020,188,851,840
	SCL_{33}	8,589,934,592	5,559,060,566,555,520
	SCL_{34}	17,179,869,184	16,677,181,699,666,600
	SCL_{35}	34,359,738,368	50,031,545,098,999,700

- Rule B: Maximum selectivity rule
  We may define the maximum number l of CO components to combine within an assembly choice of all k of CO components (note that l is max. k–1).
- Rule C: Minimum selectivity rule
  We may choose at least l components from the available possible k CO components.

In this scenario, similarly to Scenario #1, we propose two alternative groups of calculation. The first group of methods does not consider the distribution of j combinations and k combinations of components and includes two types of formulas, one for class CL_1 and another for classes $CL_{2-\infty}$.

In the case of Rule B and if $l = k$–1 and $i = 1$ (class CL_1), then the following formula can be applied:

$$\sum Conf_{CL_iSCL_j^k} = \left(\left(2^j\right)-1\right)*\left(\left(2^k\right)-1\right) \tag{6.8}$$

In the case of Rule B and if $l = k$–1 and $i = 2$–∞ (class $CL_{2-\infty}$), then the sum of the product configurations can be obtained by the formula:

$$\sum Conf_{CL_iSCL_j^k} = \left(2^j\right)*\left(\left(2^k\right)-1\right) \tag{6.9}$$

The second group of methods can be used when the distribution of both VO and CO components is a matter of interest. Then, the following two formulas can be applied:

$$\sum Conf_{CL_iSCL_j^k} = \sum_{j=1}^{n_j}\left(\frac{j!}{n_j!(j-n_j)!}\right)*\sum_{l=1}^{k}\left(\frac{k!}{l!(k-l)!}\right) \tag{6.10}$$

The formula for class $CL_{2-}\infty$ is as follows:

$$\sum Conf_{CL_iSCL_j^k} = \sum_{j=0}^{n_j}\left(\frac{j!}{n_j!(j-n_j)!}\right)*\sum_{l=1}^{k}\left(\frac{k!}{l!(k-l)!}\right) \tag{6.11}$$

An array of possible numbers of product configurations for all classes and subclasses of product configuration and for $k = <2;10>$ is depicted in Table 6.4.

Table 6.4 Fragment of table summarizing product configurations for the three selection rules (situations), while $j = 0$

Number of components (k)	Selection rule			
	(l)	Rule A	Rule B	Rule C
2	1 out of 2	2	2	2
3	1 out of 3	3	3	6
	2 out of 3	3	6	3
4	1 out of 4	4	4	14
	2 out of 4	6	10	10
	3 out of 4	4	14	4
5	1 out of 5	5	5	30
	2 out of 5	10	15	25
	3 out of 5	10	25	15
	4 out of 5	5	30	5
6	1 out of 6	6	6	62
	2 out of 6	15	21	56
	3 out of 6	20	41	41
	4 out of 6	15	56	21
	5 out of 6	6	62	6
7	1 out of 7	7	7	126
	2 out of 7	21	28	119
	3 out of 7	35	63	98
	4 out of 7	35	98	63
	5 out of 7	21	119	28
	6 out of 7	7	126	7
8	1 out of 8	8	8	254
	2 out of 8	28	36	246
	3 out of 8	56	92	218
	4 out of 8	70	162	162
	5 out of 8	56	218	92
	6 out of 8	28	246	36
	7 out of 8	8	254	8
9	1 out of 9	9	9	510
	2 out of 9	36	45	501
	3 out of 9	84	129	465
	4 out of 9	126	255	381
	5 out of 9	126	381	255
	6 out of 9	84	465	129
	7 out of 9	36	501	45
	8 out of 9	9	510	9

Table 6.4 (Continued) Fragment of table summarizing product configurations for the three selection rules (situations), while $j = 0$

Number of components (k)	Selection rule			
	(l)	Rule A	Rule B	Rule C
10	1 out of 10	10	10	1,022
	2 out of 10	45	55	1,012
	3 out of 10	120	175	967
	4 out of 10	210	385	847
	5 out of 10	252	637	637
	6 out of 10	210	847	385
	7 out of 10	120	967	175
	8 out of 10	45	1012	55
	9 out of 10	10	1022	10

The possible numbers of product configurations for an extended list of CO components $k = <11; 21>$ are available in Appendix 6.1 of this chapter.

A selection from four CO components according to Rule A follows Pascal's binomial distribution of choices. The concept of product configurations does not consider the following two situations:

- The outer left side of the triangle represents the selection

$$\binom{k}{l} = \binom{k}{0} = 1,$$

  and this would mean that no CO component must be selected. In such a case, the optional components by their nature would be invalid.
- The outer right side of the triangle (e.g., if $k = 4$) represents the selection

$$\binom{k}{l} = \binom{4}{4} = 1,$$

  and this would mean that all of the CO components must be selected. In such a case, the optional components might be replaced by the category of S components.

To present the applicability of the proposed approach, the following example for class and subclass $CL_2SCL_2^4$ is offered. It will be applied only to Rule A to present individual values of configurations per selection l out of k components. Then, by using Formula 6.11, we obtain

$$\sum Conf_{CL_2SCL_2^4} = \left[\left(\frac{2!}{0!(2-0)!}\right)+\left(\frac{2!}{1!(2-1)!}\right)+\left(\frac{2!}{2!(2-2)!}\right)\right]*\left[\left(\frac{4!}{1!(4-1)!}\right)\right]$$
$$= (1+2+1)*4 = 16$$

$$\sum Conf_{CL_2SCL_2^4} = \left[\left(\frac{2!}{0!(2-0)!}\right)+\left(\frac{2!}{1!(2-1)!}\right)+\left(\frac{2!}{2!(2-2)!}\right)\right]*\left[\left(\frac{4!}{2!(4-2)!}\right)\right]$$
$$= (1+2+1)*6 = 24$$

$$\sum Conf_{CL_2SCL_2^4} = \left[\left(\frac{2!}{0!(2-0)!}\right)+\left(\frac{2!}{1!(2-1)!}\right)+\left(\frac{2!}{2!(2-2)!}\right)\right]*\left[\left(\frac{4!}{3!(4-3)!}\right)\right]$$
$$= (1+2+1)*4 = 16$$

The product configuration generation can also be graphically demonstrated, as shown in Figure 6.4a. Alternatively, one may obtain the NPC by enumeration, as shown in Figure 6.4b.

Analogically, it is possible to apply the presented formulas to determine all possible product configurations for any subclass and class in an arbitrary individual assembly node.

6.4 Entropy-based complexity of product variety

The high frequency of topics in research publications on complexity-related problems within mass customized production clearly evokes their significance. Especially, product variety–based complexity is often a matter of interest for theorists and practitioners. Our focus in this section goes toward an exploration of variety-based complexity on the basis of AD and entropy theories. The presented approach relates to both types of product variety, internal and external.

To elucidate the basic idea of this approach to complexity, it is based on a transformation of mutual relations between input components of assembly nodes and the numbers of related product configurations into a design matrix. Subsequently, the measures to enumerate product variety complexity will be adopted.

6.4.1 Adoption of axiomatic design complexity

The four main domains of AD—customer, functional, physical, and process—were originally defined by Suh [14]. As is known, AD theory deals with satisfying customers' wishes in a manageable way, as their

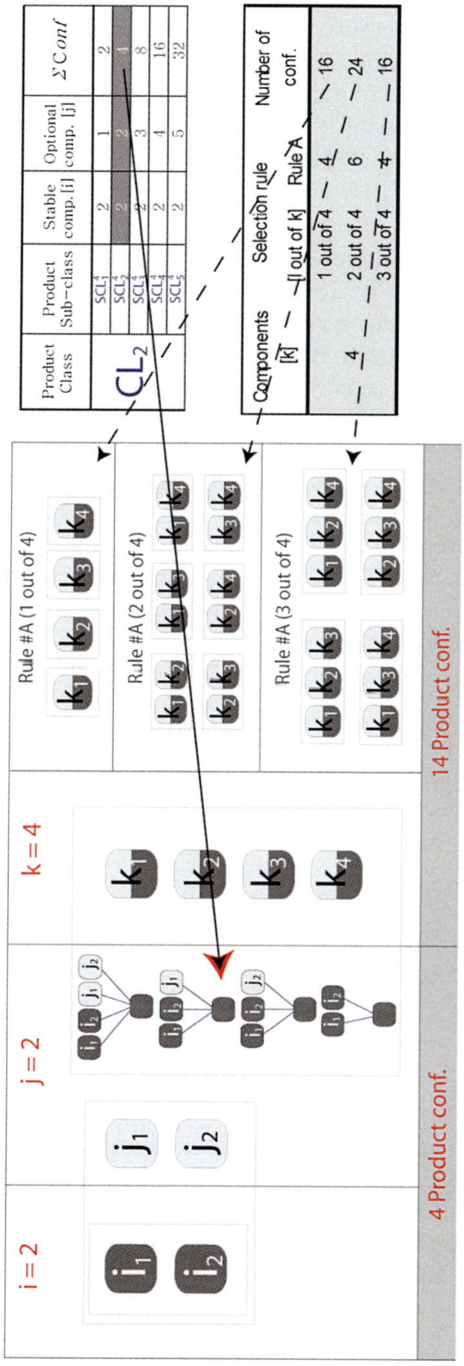

Figure 6.4 (a) Product configurations for class $CL_2SCL^4_2$ depicted graphically; (b) the number of product configurations obtained by enumeration.

wishes are often articulated in a nonengineering expression. Thus, customer needs are necessarily converted into so-called functional requirements (FRs) that must be satisfied by the design parameters (DPs). In this sense, each FR must be linked to at least one DP, according to Axiom #1 of AD theory. Then one could write

$$FR = [A]DP \tag{6.12}$$

where each FR refers to at least one coupling with DP. If the DP has no effect on FR, then the design indicates no dependency and refers to "0" and vice versa for "X," indicating coupling of DP and FR. Depending on the type of the resulting design matrix [*A*], three types of design matrices exist: uncoupled, decoupled, and coupled design.

Guenov [20] successfully tested Suh's hypothesis that the distribution of FR versus DP couplings gives a good idea of complexity. Accordingly, he derived two indicators for architectural design complexity measurement, which are relatively simple and easy to apply. A brief description of these measures follows. Let us denote N as the number of interactions within a design matrix, and N_1, N_2,..., N_K as the numbers of interactions per each DP of the same matrix. Then, the so-called degree of disorder W can be expressed by the formula:

$$\Omega = C_N^{N_1} * C_{N-N_1}^{N_2} * C_{N-N_1-N_2}^{N_3} \ldots * \ldots C_{N-N_1-\ldots-N_{K-1}}^{N_K} = \frac{N!}{N_1!N_2!\ldots N_K!} \tag{6.13}$$

where:

$$C_N^{N_1} = \frac{N!}{(N-N_1)!N_1!} \tag{6.14}$$

Secondly, systems design complexity (SDC), as denoted by the authors, is expressed as follows:

$$SDC = \sum N_P \, ln \, N_P \tag{6.15}$$

where N_P is interpreted as the number of interactions per single DP.

Based on our previous works [21] where comparison of the two indicators was performed, it was recognized that the SDC indicator is more suitable for the given application domain than the degree of disorder.

In order to adopt the SDC indicator in terms of MC, it is firstly important to transform our concept for determining product configurations into the design matrix [*A*]. It is advisable to adopt the model for the generation of all possible product configurations and their bipartite

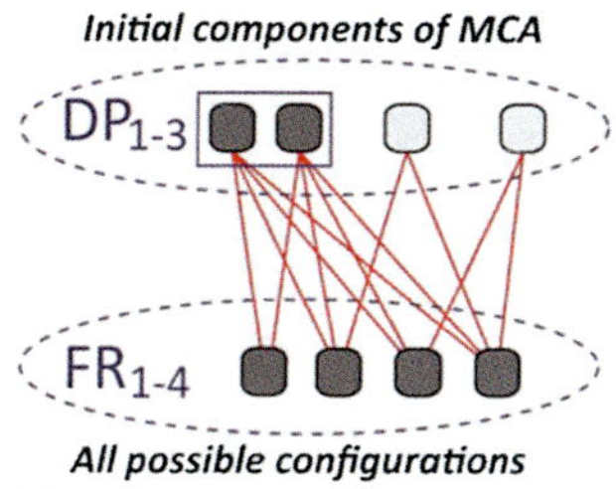

(a)

	DP_1	DP_2	DP_3
FR_1	*X*		
FR_2	*X*	*X*	
FR_3	*X*		*X*
FR_4	*X*	*X*	*X*

(b)

Figure 6.5 (a) Case model of MCA as a bipartite graph; (b) AD matrix of the case model.

graphical representation into AD matrices under the precondition that FRs will be represented by all possible product configurations and DPs will be represented by initial components related to product configurations (Figure 6.5a).

Based on that, it is possible to transform practically any model of initial components and related product configurations into an AD matrix, as shown in Figure 6.5b. Subsequently, one can use the obtained design matrices to apply the complexity measures described by Formula 6.13 and/or Formula 6.15.

6.4.2 *Comparison of axiomatic design-based and combinatorial product complexity indicators*

For this purpose, three testing cases will be considered for individual selections from four to six CO components. Two indicators will be benchmarked—namely, the NPC and the SDC. The obtained values are depicted in Table 6.5.

As can be seen from Table 6.5, the values of the NPC indicator for individual selections from four to six CO components are in line with existing combinatorial phenomenon on the distribution of possible combinations within Pascal's triangle. With regard to the values of the SDC indicator in all three subclasses $CL_{1-\infty}SCL_1^4$, $CL_{1-\infty}SCL_1^5$, and $CL_{1-\infty}SCL_1^6$, these do not follow the symmetrical values and are unique to every individual selection rule. Moreover, it can be easily proved that, for example, choosing one out of six options is less complex than choosing five out of six options. The product variety complexity for the "one out of six" selection, then, is lower than in the other cases of this subclass. The difference is a consequence of the higher number of bipartite interactions, so that

$$\sum N_{p\binom{6}{1}} = 30 < \sum N_{p\binom{6}{5}} = 78.$$

Table 6.5 Fragment of complexity indicators NPC and SDC

i	j	k	Selection rule	NPC	SN_p	SDC
1–∞	1	4	1 out of 4	8	20	27,7
			2 out of 4	12	42	83,6
			3 out of 4	8	36	65,2
		5	1 out of 5	10	25	38
			2 out of 5	20	70	166,1
			3 out of 5	20	90	232
			4 out of 5	10	50	114,3
		6	1 out of 6	12	30	48,9
			2 out of 6	30	105	280,8
			3 out of 6	40	180	567
			4 out of 6	30	165	502,1
			5 out of 6	12	78	178,7

Thus, an important finding is that the SDC indicator shows the most realistic complexity values compared with the number of possible product configurations.

6.5 *Feed and transfer complexity conception*

The product variety complexity measures described in Section 6.4 have been developed with the purpose of quantifying complexity for individual assembly nodes. Their application would be very limited since real assembly processes consist of multiple layers and complicated networks. Unlike SDC, the configuration complexity (CC) expressed by all possible product configurations ($\Sigma Conf$) is easily applicable even for complex modular structures.

Further, an effective way to apply these measures in cases of real assembly processes will be described. According to Hu et al. [22], the complexity of individual assembly stations is obtained as a weighed sum of complexities associated with all upstream assembly activities, as can be seen in Figure 6.6.

Feed complexity exists due to the product configurations added onto the previous stations, and they affect subsequent processes at stations and configuration selections.

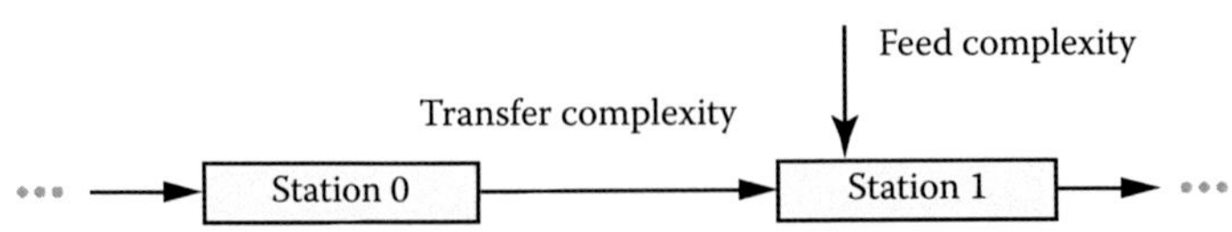

Figure 6.6 Complexity aggregation principle.

According to the previous scheme, transfer complexity can only flow from upstream to downstream, not in the opposite direction. So-called feed complexity can only be added at a current station without any transferring behavior. Then, the total complexity is always the sum of feed/node complexity and transfer complexity from all upstream assembly stations. These aggregation and/or cumulative principles are further applied to calculate the total complexity of modular structures.

6.6 *Case application of entropy-based and combinatorial product configuration complexities*

To show the relevance of the aforementioned methodological frameworks, two selected measures will be applied and verified on a model of the mass customized manufacturing of washing machines (Figure 6.7). The assembly process consists of two independent branches. Branch #1 offers the top-loading version of a washing machine and Branch #2 is dedicated to the front-loading option. Customized assembly branches on the basis of two determining S input components: A_1 for the top-loading and A_2 for the front-loading machine option. The second type is from voluntary components $B_{1–3}$, $C_{1–4}$, $D_{1–4}$, $E_{1–3}$, $H_{1–2}$, and $I_{1–2}$, which offer so-called standard options. This option is automatically selected in cases when any of the nonstandard options is selected. The last component type—CO $F_{1–4}$, $G_{1–5}$—are those with an obligation to choose at least one component option of all possible within the appropriate assembly module (selection Rule C).

Figure 6.7 presents summary values of all possible product configurations in individual modules and the total number of product configurations.

From Figure 6.8 it is clear that if a specific node on the single level joins multiple components/modules and these modules are dependent, then the summary value of $\sum Conf$ is a multiplication of the incoming (converging) product configurations on the same level. In the case of independent branches (e.g., Branch #1 and Branch #2 on tier t_0 in Figure 6.8), then the summary value of $\sum Conf$ for such branches is a sum of the incoming (converging) product configurations into the node.

In Figure 6.8, aggregated complexity values based on the AD approach using Formula 6.15 are presented summarily.

Summary values of the SDC complexity for both dependent and independent branches are obtained as a summation of the feed (node) complexity and complexity transferred from previous upstream stations. As can be seen in Figure 6.8, the summary SDC value on the final assembly station (5, with independent branches) is obtained as a sum of the two converging complexities, while feed SDC on station 5 is zero, due to the absence of additional operations.

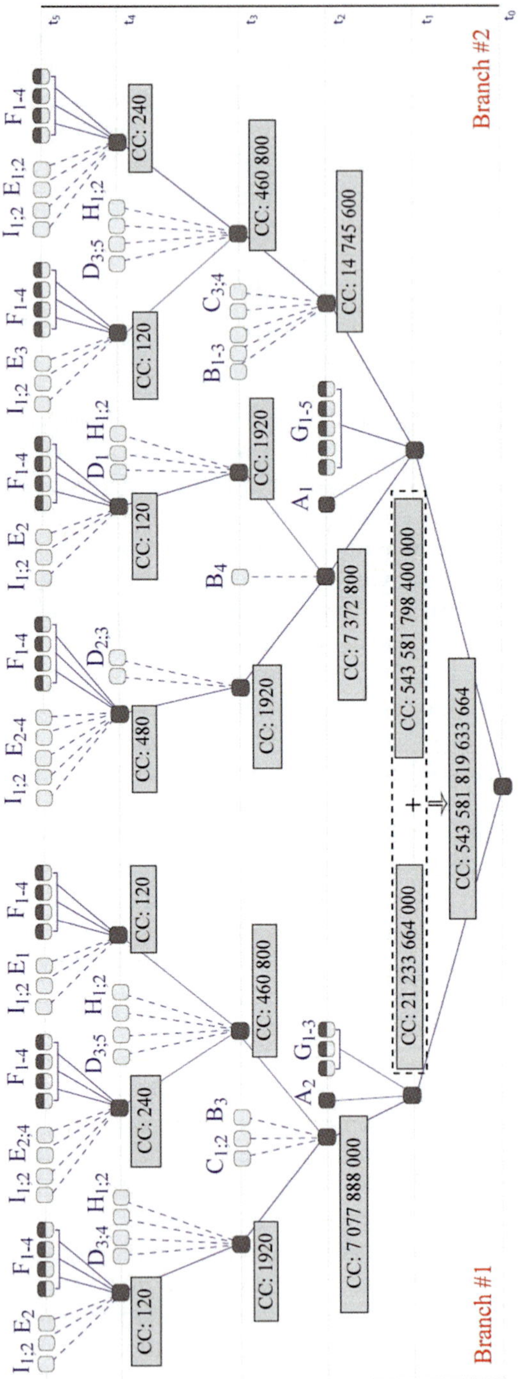

Figure 6.7 Model of MCA with CC value for individual nodes and summary CC value, including legend.

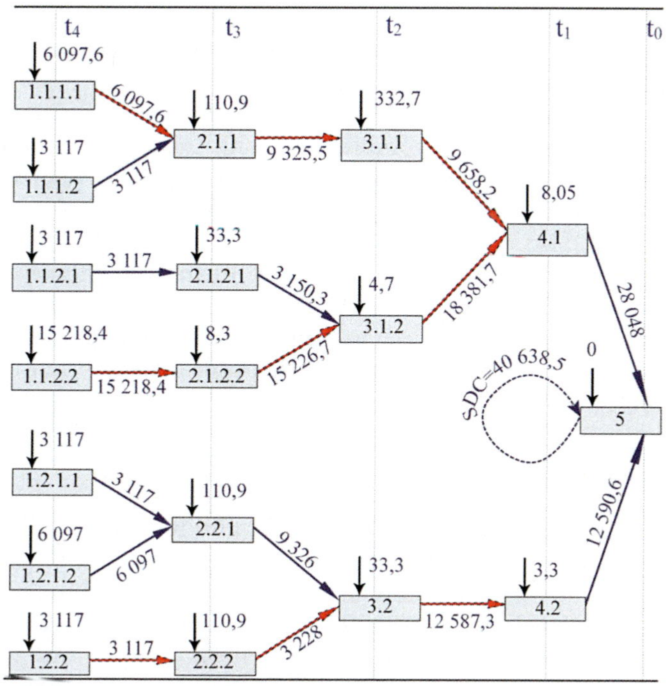

Figure 6.8 Summary feed SDC values of the MCA model.

In summary, regarding so-called flow complexity, the following can be stated.

- In *independent branches* that result in a *final product,* final complexity in both approaches is obtained as a sum of individual complexities.
- In *dependent branches* resulting in *semiproducts,* the CC of the next downstream node is obtained as a multiplication of converging complexities, which is moreover multiplied by the optional component(s) according to combinatorial rules.

6.7 Conclusion

This chapter considered a number of approaches to assessing product variety complexity. One of the approaches—namely, the combinatorial-based approach to quantifying the NPC in MC—is quite straightforward, thus product development teams can effectively measure the variety of concurrent product platforms. Counting the number of product variations, as in the second method described in Section 6.2.2, seems not to be very efficient, as product variations represent only imaginary products connecting all voluntary components in different positions of S components.

The measure adopting AD entropy, called SDC, proved that this indicator works better for the given purpose. This has been shown in the case application in Section 6.6.

An assessment of the product variety complexity of modular models by the SDC indicator in combination with the aggregation principle results in flat values of complexity. Vice versa, complexity values obtained by NPC indicators are generated as a multiplication of all converging complexities and bring exponential values.

Finally, both measures (NPC and SDC) can be combined to assist product managers and marketing advisors to independently assess alternative product platforms and to evaluate their customization characteristics, such as product variety complexity, in order to decide on the optimal customizable product platforms.

References

1. Tiihonen, J., Soininen, T., Männistö, T., and Sulonen, R. (1996). State-of-the-practice in product configuration: A survey of 10 cases in the Finnish industry. In T. Tomiyama, M. Mantyla, and S. Finger, Eds., *Knowledge Intensive CAD*, pp. 95–114. Springer, New York.
2. Klein, R., Buchheit, M., and Nutt, W. (1994). Configuration as model construction: The constructive problem solving approach. In J. S. Gero, and F. Sudweeks, Dordrecht, *Proceedings of Artificial Intelligence in Design 1994*, pp. 201–218, Springer, Netherlands.
3. Mittal, S., and Frayman, F. (1989). Towards a generic model of configuration tasks. In N. S. Sridharan Ed., *IJCAI-89: Proceedings of the Eleventh IJCAI*, pp. 20–25, Morgan Kaufman, San Francisco.
4. Najman, O., and Stein, B. (1992). A theoretical framework for configurations. In F. Belli and F. J. Radermacher, Eds., *Proceedings of Industrial and Engineering Applications of Artificial Intelligence and Expert Systems: 5th International Conference, IEA/AIE-92*, pp. 441–450, Springer Verlag, Berlin.
5. Calinescu, A., Efstathiou, J., Schirn, J., and Bermejo, J. (1998). Applying and assessing two methods for measuring complexity in manufacturing. *Journal of the Operational Research Society*, 49(7), 723–733.
6. Jiao, J., Ma, Q., and Tseng, M. M. (2003). Towards high value-added products and services: Mass customization and beyond. *Technovation*, 23(10), 809–821.
7. Liu, X. F., Kane, G., and Bambroo, M. (2006). An intelligent early warning system for software quality improvement and project management. *Journal of Systems and Software*, 79(11), 1552–1564.
8. Fredendall, D., and Gabriel, T. J. (2003). Manufacturing complexity: A quantitative measure. *POMS Conference: Proceedings*, April 4–7, 2003, Savannah, GA.

9. Shannon, C. E. (1948). A mathematical theory of communication. *Bell System Technical Journal*, 27(3), 379–423.
10. Zhu, X., Koren, S. J., and Marin, S. P. (2008). Modeling of manufacturing complexity in mixed-model assembly lines. *Journal of Manufacturing Science and Engineering: Transactions of the ASME*, 130(5), 313–334.
11. Desmukh, A. V., Talavage, J. J., and Barash, M. M. (1998). Complexity in manufacturing systems, Part 1: Analysis of static complexity. *IEEE Transactions on Engineering Management*, 30(7), 645–655.
12. Suh, N. P. (2005). Complexity in engineering. *CIRP Annals: Manufacturing Technology*, 54(2), 46–63.
13. Kim, Y.-S. (1999). A system complexity approach for the integration of product development and production system design. MSc thesis, MIT, Cambridge, MA.
14. Frizelle, G., and Woodcock, E. (1995). Measuring complexity as an aid to developing operational strategy. *International Journal of Operations and Production Management*, 15, 26–39.
15. Frizelle, G., and Efstathiou, J. (2002). Seminar notes on measuring complex systems. Resource document. The London School of Economics and Political Science. Retrieved from http://www.psych.lse.ac.ukcomplexity/PDFiles/Seminars/GerjanApril02lastversion.pdf. Accessed May 12, 2014.
16. Krus, P. (2015). Design space configuration for minimizing design information entropy. In A. Chakrabarti Ed., *Proceedings of the ICoRD '15: Research into Design across Boundaries; Theory, Research Methodology, Aesthetics, Human Factors and Education*, pp. 51–60, Springer, New Delhi, India.
17. Modrak, V., Krus, P., and Bednar, S. (2015). Approaches to product variety management assuming configuration conflict problem. *FME Transaction*, 43, 271–278.
18. Kampker, A., Burggräf, P., Swist, M., and Nowacki, C. (2014). Assessment and configuration of a product production system. In H. ElMaraghy Ed., *Enabling Manufacturing Competitiveness and Economic Sustainability*, pp. 147–152, Springer, Berlin.
19. Grussenmeyer, R., and Blecker, T. (2013). Requirements for the design of a complexity management method in new product development of integral and modular products. *International Journal of Engineering, Science and Technology*, 5(2), 132–149.
20. Guenov, M. D. (2002). Complexity and cost-effectiveness measures for systems design. In G. Frizelle, and H. Richards Eds., *Manufacturing Complexity Network Conference 2002*, pp. 455–465, Cambridge, UK.
21. Modrak, V., and Bednar, S. (2015). Using axiomatic design and entropy to measure complexity in mass customization. *Procedia CIRP*, 34, 87–92.
22. Hu, S. J., Zhu, X., Wang, H., and Koren, Y. (2008). Product variety and manufacturing complexity in assembly systems and supply chains. *Annals of CIRP*, 57, 45–48.

Appendix 6.1

Summary table of available product configurations when $k = 11–21$

			Selection rules		
Components (*k*)	(*l*)	(*l* out of *k*)	Rule A	Rule B	Rule C
11	1	1 out of 11	11	11	2,047
	2	2 out of 11	55	66	2,036
	3	3 out of 11	165	231	1,981
	4	4 out of 11	330	561	1,816
	5	5 out of 11	462	1,023	1,486
	6	6 out of 11	462	1,485	1,024
	7	7 out of 11	330	1,815	562
	8	8 out of 11	165	1,980	232
	9	9 out of 11	55	2,035	67
	10	10 out of 11	11	2,046	12
12	1	1 out of 12	12	12	4,095
	2	2 out of 12	66	78	4,083
	3	3 out of 12	220	298	4,017
	4	4 out of 12	495	793	3,797
	5	5 out of 12	792	1,585	3,302
	6	6 out of 12	924	2,509	2,510
	7	7 out of 12	792	3,301	1,586
	8	8 out of 12	495	3,796	794
	9	9 out of 12	220	4,016	299
	10	10 out of 12	66	4,082	79
	11	11 out of 12	12	4,094	13
13	1	1 out of 13	13	13	8,191
	2	2 out of 13	78	91	8,178
	3	3 out of 13	286	377	8,100
	4	4 out of 13	715	1,092	7,814
	5	5 out of 13	1,287	2,379	7,099
	6	6 out of 13	1,716	4,095	5,812
	7	7 out of 13	1,716	5,811	4,096
	8	8 out of 13	1,287	7,098	2,380
	9	9 out of 13	715	7,813	1,093
	10	10 out of 13	286	8,099	378
	11	11 out of 13	78	8,177	92
	12	12 out of 13	13	8,190	14
14	1	1 out of 14	14	14	16,383
	2	2 out of 14	91	105	16,369
	3	3 out of 14	364	469	16,278

Components (*k*)	(*l*)	(*l* out of *k*)	Selection rules Rule A	 Rule B	 Rule C
	4	4 out of 14	1,001	1,470	15,914
	5	5 out of 14	2,002	3,472	14,913
	6	6 out of 14	3,003	6,475	12,911
	7	7 out of 14	3,432	9,907	9,908
	8	8 out of 14	3,003	12,910	6,476
	9	9 out of 14	2,002	14,912	3,473
	10	10 out of 14	1,001	15,913	1,471
	11	11 out of 14	364	16,277	470
	12	12 out of 14	91	16,368	106
	13	13 out of 14	14	16,382	15
15	1	1 out of 15	15	15	32,767
	2	2 out of 15	105	120	32,752
	3	3 out of 15	455	575	32,646
	4	4 out of 15	1,365	1,940	32,192
	5	5 out of 15	3,003	4,943	30,827
	6	6 out of 15	5,005	9,948	27,824
	7	7 out of 15	6,435	16,383	22,819
	8	8 out of 15	6,435	22,818	16,384
	9	9 out of 15	5,005	27,823	9,949
	10	10 out of 15	3,003	30,826	4,944
	11	11 out of 15	1,365	32,191	1,941
	12	12 out of 15	455	32,646	576
	13	13 out of 15	105	32,751	121
	14	14 out of 15	15	32,766	16
16	1	1 out of 16	16	16	65,535
	2	2 out of 16	120	136	65,519
	3	3 out of 16	560	696	65,399
	4	4 out of 16	1,820	2,516	64,839
	5	5 out of 16	4,368	6,884	63,019
	6	6 out of 16	8,008	14,892	58,651
	7	7 out of 16	11,440	26,332	50,643
	8	8 out of 16	12,870	39,202	39,203
	9	9 out of 16	11,440	50,642	26,333
	10	10 out of 16	8,008	58,650	14,893
	11	11 out of 16	4,368	63,018	6,885
	12	12 out of 16	1,820	64,838	2,517
	13	13 out of 16	560	65,398	697
	14	14 out of 16	120	65,518	137

(*Continued*)

			Selection rules		
Components (k)	(l)	(l out of k)	Rule A	Rule B	Rule C
	15	15 out of 16	16	65,534	17
17	1	1 out of 17	17	17	131,071
	2	2 out of 17	136	153	131,054
	3	3 out of 17	680	833	130,918
	4	4 out of 17	2,380	3,213	130,238
	5	5 out of 17	6,188	9,401	127,858
	6	6 out of 17	12,376	21,777	121,670
	7	7 out of 17	19,448	41,225	109,294
	8	8 out of 17	24,310	65,535	89,846
	9	9 out of 17	24,310	89,845	65,536
	10	10 out of 17	19,448	109,293	41,226
	11	11 out of 17	12,376	121,669	21,778
	12	12 out of 17	6,188	127,857	9,402
	13	13 out of 17	2,380	130,237	3,214
	14	14 out of 17	680	130,917	834
	15	15 out of 17	136	131,053	154
	16	16 out of 17	17	131,070	18
18	1	1 out of 18	18	18	262,143
	2	2 out of 18	153	171	262,125
	3	3 out of 18	816	987	261,972
	4	4 out of 18	3,060	4,047	261,156
	5	5 out of 18	8,568	12,615	258,096
	6	6 out of 18	18,564	31,179	249,528
	7	7 out of 18	31,824	63,003	230,964
	8	8 out of 18	43,758	106,761	199,140
	9	9 out of 18	48,620	155,381	155,382
	10	10 out of 18	43,758	199,139	106,762
	11	11 out of 18	31,824	230,963	63,003
	12	12 out of 18	18,564	249,527	31,180
	13	13 out of 18	8,568	258,095	12,616
	14	14 out of 18	3,060	261,155	4,048
	15	15 out of 18	816	261,971	988
	16	16 out of 18	153	262,124	172
	17	17 out of 18	18	262,142	19
19	1	1 out of 19	19	19	524,287
	2	2 out of 19	171	190	524,268
	3	3 out of 19	969	1,159	524,097
	4	4 out of 19	3,876	5,035	523,128
	5	5 out of 19	11,628	16,663	519,252

			Selection rules		
Components (*k*)	(*l*)	(*l* out of *k*)	Rule A	Rule B	Rule C
	6	6 out of 19	27,132	43,795	507,624
	7	7 out of 19	50,388	94,183	480,492
	8	8 out of 19	75,582	169,765	430,104
	9	9 out of 19	92,378	262,143	354,522
	10	10 out of 19	92,378	354,521	262,144
	11	11 out of 19	75,582	430,103	169,766
	12	12 out of 19	50,388	480,491	94,184
	13	13 out of 19	27,132	507,623	43,796
	14	14 out of 19	11,628	519,251	16,664
	15	15 out of 19	3,876	523,127	5,036
	16	16 out of 19	969	524,096	1,160
	17	17 out of 19	171	524,267	191
	18	18 out of 19	19	524,286	20
20	1	1 out of 20	20	20	1,048,575
	2	2 out of 20	190	210	1,048,555
	3	3 out of 20	1,140	1,350	1,048,365
	4	4 out of 20	4,845	6,195	1,047,225
	5	5 out of 20	15,504	21,699	1,042,380
	6	6 out of 20	38,760	60,459	1,026,876
	7	7 out of 20	77,520	137,979	988,116
	8	8 out of 20	125,970	263,949	910,596
	9	9 out of 20	167,960	431,909	784,626
	10	10 out of 20	184,756	616,665	616,666
	11	11 out of 20	167,960	784,625	431,910
	12	12 out of 20	125,970	910,595	263,950
	13	13 out of 20	77,520	988,115	137,980
	14	14 out of 20	38,760	1,026,875	60,460
	15	15 out of 20	15,504	1,042,379	21,700
	16	16 out of 20	4,845	1,047,224	6,196
	17	17 out of 20	1,140	1,048,364	1,351
	18	18 out of 20	190	1,048,554	211
	19	19 out of 20	20	1,048,574	21
21	1	1 out of 21	21	21	2,097,151
	2	2 out of 21	210	231	2,097,130
	3	3 out of 21	1,330	1,561	2,096,920
	4	4 out of 21	5,985	7,546	2,095,590
	5	5 out of 21	20,349	27,895	2,089,605
	6	6 out of 21	54,264	82,159	2,069,256

(*Continued*)

			Selection rules		
Components (k)	(l)	(l out of k)	Rule A	Rule B	Rule C
	7	7 out of 21	116,280	198,439	2,014,992
	8	8 out of 21	203,490	401,929	1,898,712
	9	9 out of 21	293,930	695,859	1,695,222
	10	10 out of 21	352,716	1,048,575	1,401,292
	11	11 out of 21	352,716	1,401,291	1,048,576
	12	12 out of 21	293,930	1,695,221	695,860
	13	13 out of 21	203,490	1,898,711	401,930
	14	14 out of 21	116,280	2,014,991	198,440
	15	15 out of 21	54,264	2,069,255	82,160
	16	16 out of 21	20,349	2,089,604	27,896
	17	17 out of 21	5,985	2,095,589	7,547
	18	18 out of 21	1,330	2,096,919	1,562
	19	19 out of 21	210	2,097,129	232
	20	20 out of 21	21	2,097,150	22

chapter seven

Product variety management assuming product configuration conflicts

Vladimir Modrak and Slavomir Bednar

Contents

ABSTRACT
An important part of product variety management is finding the optimum variety extent. Usually, the product variety extent is limited by, for example, production capabilities and/or production capacities, among other restrictions. This chapter intends to show that the product variety extent can also be influenced by configuration conflicts. Making decisions about product configurations is part of standard product architecture development activity. Producers need to deliver variants to address the diversity of consumer needs. However, when configuration conflicts occur, the question is how they will be perceived by customers. Here, we propose an analytical method to be used in decisions about the elimination or retainment

of configuration conflicts. This method will be verified through a realistic case where alternative product design platforms are compared. The newly developed method can be employed to assist product managers to independently assess competitive product variety platforms against each other.

7.1 Introduction

Once managers are in the early stages of product architecture design, they might decide about the most suitable product component/module structure. Normally, marketing managers strive to maximize the variety on offer with the aim to satisfy a wide range of customers, knowing also that some incompatible components can occur in possible product configurations. The problem is that they are not aware of the number of infeasible product configurations. Moreover, it is not easy to identify them using calculation methods. Anyhow, a relatively high number of such infeasible product configurations, as a rule, negatively affects customer perception and buying behavior. On the other hand, if these infeasible product configurations were to be totally eliminated, it would have a negative impact on the extent of customization. Normally, the extent of customization for products is perceived in the sense that the bigger the product variety, the better, and vice versa [1]. Therefore, the extent of customization required by customers should not be ignored to remain competitive.

However, high product variety is resulting in less flexibility and higher costs in manufacturing systems and is becoming a serious problem. Product designers, then, have to consider both problems and find an optimal balance between them.

The main scope of this chapter is to explore the possibilities of solving this issue by changing the rate between infeasible product configurations and all possible product configurations when restrictions are omitted. In a simple way, the number of product configurations is closely related to variety-induced complexity. However, the number of product configurations, both viable as well as unviable, is not an optimal indicator of the variety-induced complexity used to solve this problem and to express the rate. Therefore, instead of the number of product configurations, entropy-based complexity metrics will be used as a tool for decision-making in product variety management. Finally, in this chapter, a decision-making algorithm to solve issues related to the optimal selection of a product component platform will be presented.

7.2 Entropy-based complexity metrics for product variety extent

7.2.1 Theoretical background

The aim of this subsection is to analyze the relation between infeasible product configurations and all possible product configurations. The very first notion of complexity was outlined in the work of Shannon [2], where information theory was originally introduced. A few years later, information became a key complexity element for the description and analysis of systems entropy. Discrete-case entropy (H_d) has been defined by the probability P_i of the n-state occurrence, as follows [2]:

$$H_d = \sum_{i=1}^{n} P_i \log_2 P_i \tag{7.1}$$

Differential information entropy of the probability density function $p(x)$ for continuous signals (H_c) has been expressed as

$$H_c = \int_{-\infty}^{\infty} p(x)\log_2\left(p(x)\right)dx \tag{7.2}$$

Krus [3] adopted design information entropy for the multidimensional case (H_x) in the following form:

$$H_x = \int_{-\infty} p(x)\log_2\left(p(x)S\right)dx \tag{7.3}$$

where D is the design space where the particular design x is defined. S is the size of the design space, expressed as

$$S = \int_{D} x\,dx \tag{7.4}$$

In the case of a general multivariable design space, its information entropy H can be expressed as follows:

$$H = \log_2 \frac{S}{s} \tag{7.5}$$

where s is the region of uncertainty for the final design of the validated system architecture.

According to Krus [3], each particular design x with regard to its design space has information entropy H_x:

$$H_x = \log_2 n_s \tag{7.6}$$

where n_s is the number of unique design alternatives (representing the so-called complete design space) that are the result of a combination of product options and H_x is denoted as the entropy of the complete design space.

There are many real cases in which some variants are impractical due to the presence of constraint(s). The information entropy, then, of constrained design space H_c can be enumerated as

$$H_c = \log_2 n_v \tag{7.7}$$

where n_v is a number of viable design alternatives.

Since a higher number of all possible design variants has a more positive impact on consumers than a smaller, constrained design space, the entropy of the constrained design space in terms of the mass customization (MC) environment should be maximized. In this sense, the entropy of a constrained design space can be considered positive entropy.

In this context, the same author expressed the quality of a modular design through the rest of the design space outside the constrained space with the term *waste information entropy* and quantified it using the formula:

$$H_W = H_X - H_C \tag{7.8}$$

In line with the logic used for the characteristics of constrained design space entropy, waste entropy can be considered negative.

Once the background of waste entropy is outlined, we may proceed toward its application.

In order to catch the effect of product design optimization by using the concept of negative entropy, we firstly need to generate concurrent product design architectures that will be mutually benchmarked. One way to do so is through the gradual execution of selected components from an original product design architecture. Subsequently, the mutual relation between so-called positive and negative entropy can be purposefully analyzed. For this objective, it is necessary to enumerate the number of all possible product configurations when restrictions are omitted, and all possible product configurations with component restrictions. This procedure is shown in the following section.

7.2.2 *Methodical approach to enumerating product configurations and waste entropy*

To show the practicability of this approach, a realistic case is provided to motivate practitioners to solve possible similar problems. For this reason, an assembly model of a personal computer adopted from Yang and Dong [4] has been used to identify product configurations, as seen in Figure 7.1.

At a higher degree of customization, there may be reasons for restrictions and/or obligations between two or more components. They may be of a functional, design, connectivity, or other nature. In order to solve the possible consequences of restrictions and obligations, the required relations or incompatibilities between components might be specified for any product architecture. In our experimental assembly case, four types of configuration rules may arise [4]:

- Require rule
- Incompatible rule
- Port connection rule
- Resource balancing rule

The model depicted in Figure 7.1 is a representation of the MC assembly of a personal computer consisting of five basic modules: a CD drive (one option), an HD unit (six individual customer options), a motherboard (MB) (three options: MB1, MB2, and MB3), a CPU (586I, 586II, 486), and a server operating system (OS) (OS1 and OS2). The case model has various customizable options depending on the customer choice but with predefined restrictions in the form of rules related to the incompatibility of components, defined as follows:

R#1: Component CPU3 must not be in the same configuration as component MB1.
R#2: Component MB2 must not be in the same configuration as components CPU1 and CPU2.
R#3: Component CPU3 must not be in the same configuration as component MB3.
R#4: Component OS1 must not be in the same configuration as components MB1 and MB3.
R#5: Component OS2 must not be in the same configuration as components MB2 and MB3.
R#6: Components MB2 and MB3 must not be in the same configuration as components HD4, HD5, and HD6.
R#7: Component OS2 must not be in the same configuration as components HD2 and HD4.

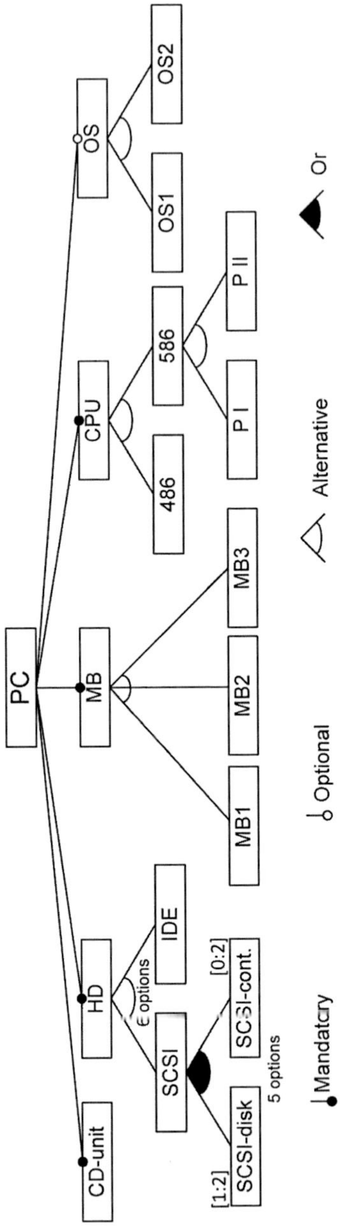

Figure 7.1 Product structure with rules for constraint satisfaction problems.

7.2.2.1 *Enumeration of product configurations with and without restrictions*

At the beginning, it is useful to transform the computer structure with the constraints shown in Figure 7.1 into a simplified assembly graph, depicted in Figure 7.2. Any such structure usually consists of a number of assembly stations (*nodes*). These can be identified within a multilevel network. In our case, a two-level network is sufficient to model the final assembly operation of the personal computer. Additionally, a specific number of component alternatives can be identified at each node of tier t_1.

The HD unit is represented by six individual alternatives. The number of all possible component combinations is seven, but one of them is omitted—namely, the combination consisting of two SCSI controllers with a single SCSI disk, as the second controller in such a hypothetical product is considered redundant.

On the bottom tier (t_0) all possible product configurations without restrictions can be identified for the original product design platform (D_0):

$$\sum Conf_{D_0} = 1*6*3*3*2 = 108$$

Subsequently, it is necessary to determine the total number of configurations when restriction rules are considered. For this purpose, an incidence matrix with component restrictions R#1–7 is constructed (see Table 7.1).

To enumerate the number of restricted product configurations, the following procedure is proposed. In the first step, let us select, for example, a group of HD units. Then, we select an arbitrary configuration from the group—for example, HD2, which is one of the six HD unit options. Afterward, we may construct an incidence submatrix for the HD2 option and the group of CPU components. As there are no restrictions, HD2 as an option can be combined with any CPU component (see Figure 7.3). Then,

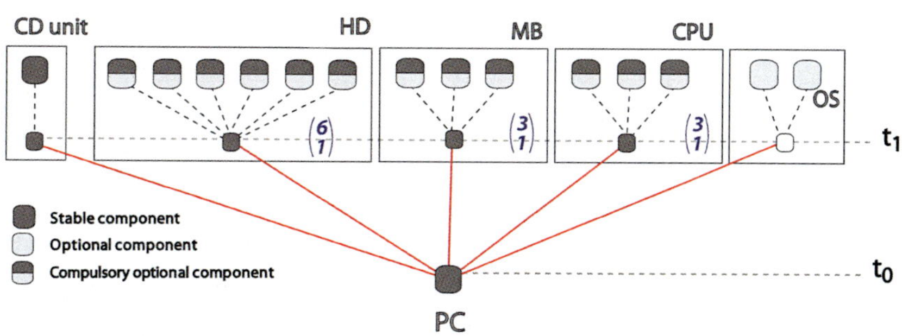

Figure 7.2 Assembly graph of a personal computer without component restrictions.

Table 7.1 Incidence matrix for component restrictions *R*#1–7

		Group 1 (HD)						Group 2 (CPU)			Group 3 (MB)			Group 4 (OS)	
		HD1	HD2	HD3	HD4	HD5	HD6	CPU1	CPU2	CPU3	MB1	MB2	MB3	OS1	OS2
Group 1 (HD)	HD1														
	HD2														❼
	HD3														
	HD4											❻	❻		❼
	HD5											❻	❻		
	HD6											❻	❻		
Group 2 (CPU)	CPU1											❷			
	CPU2											❷			
	CPU3										❶		❸		
Group 3 (MB)	MB1													❹	
	MB2														❺
	MB3													❹	❺
Group 4 (OS)	OS1														
	OS2														

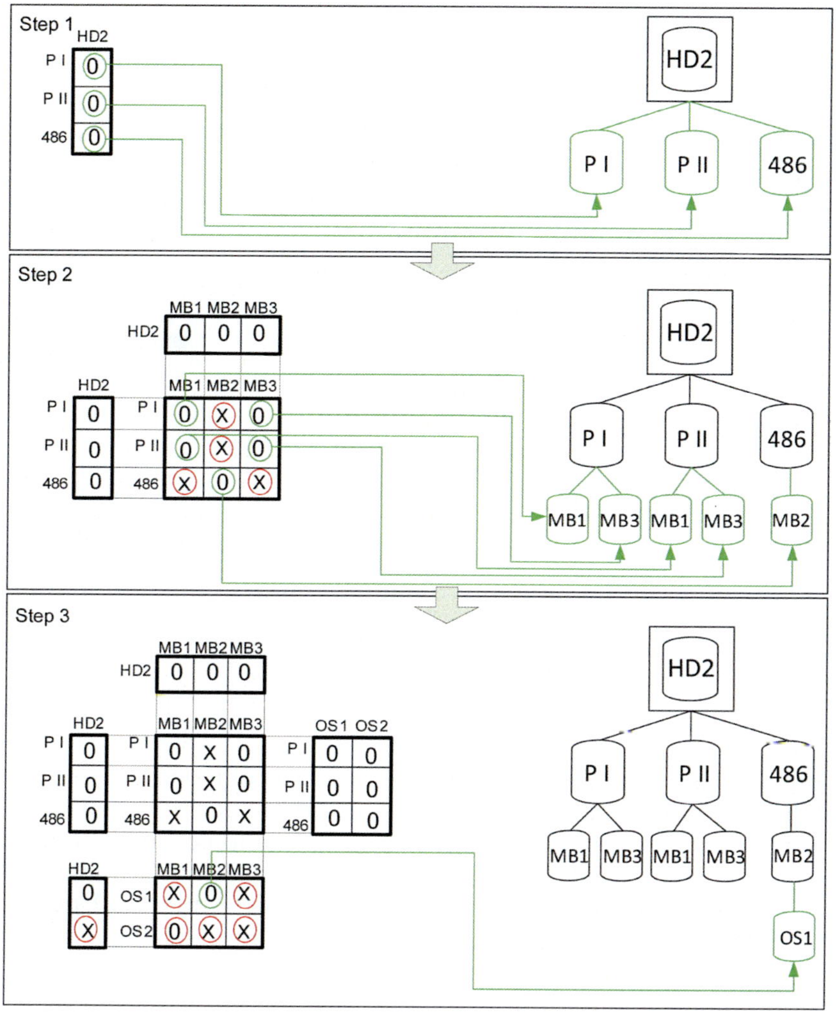

Figure 7.3 Proposed approach to transforming an incidence matrix into a product configurations model.

we need to create a three-dimensional matrix of relations between the configurations of HD2, the group of CPU components, and the group of motherboard components (Step 2). Four restrictions are identified and, accordingly, CPU components can be combined with compatible motherboard components (see Figure 7.3). Finally, four-dimensional matrix relations are constructed. Then, it is possible to exactly determine the number of restricted product configurations where HD2 is exclusively involved.

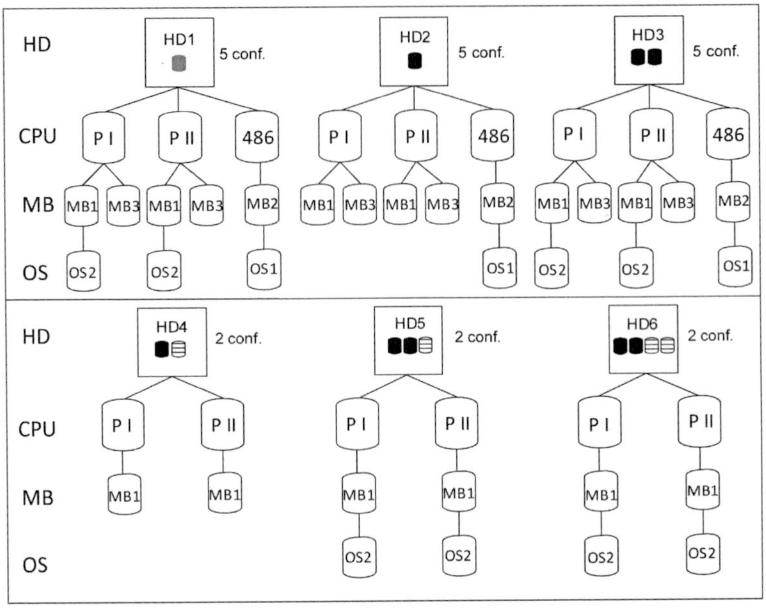

Figure 7.4 Model of 21 possible product configurations respecting the configuration rules.

Moreover, this procedure generates the product component structure of all the identified product configurations, as can be seen in Figure 7.3.

The subprocedure depicted in Figure 7.3 has to be repeated for the rest of the components from group 1. The sum of configurations for the individual components of group 1, then, determines all possible restricted product configurations. The sum is 21, as can be seen from Figure 7.4.

7.2.2.2 Proposed procedure to reduce waste entropy

As mentioned in the introduction, the goal of the configuration conflict solutions in terms of MC is to reduce the number of infeasible configurations. One possible way to reach this goal is changing the rate between infeasible product configurations and all possible product configurations when restrictions are omitted [5]. This rate can be changed through the execution of restricted components from an original product design platform (D_0).

For this reason, a new product design platform (D_1) can be obtained when, for example, one of the motherboards (namely, MB2) is selected for execution. Configurations where MB2 is present are not counted, and therefore, the total number of model configurations equals 72 without accepting rules and restrictions R#1–7. The number was reached by the following multiplication:

$$\sum Conf_{D_1} = 1*6*2*3*2 = 72$$

Then, applying the procedure proposed in Figure 7.3, the number of available restricted product configurations will decrease to 18, as enumerated by the following formula:

$$\sum Res\_Conf_{D_1} = 5+5+5+2+2+2 = 18$$

To obtain another alternative product design structure (D_2) for benchmarking purposes, another component (CPU3) has been eliminated. Then, the number of total model configurations is calculated as follows:

$$\sum Conf_{D_2} = 1*6*2*2*2 = 48$$

The number of restricted product configurations will remain at 18. The obtained numbers of configurations with and without restrictions are summarily depicted in Table 7.2.

Subsequently, waste entropy and the waste entropy rate for each of the design platforms D_{0-2} is calculated in Table 7.3.

Table 7.3 shows how the waste entropy ratio has changed by reducing the number of restricted components. Both the reductions from D_0 to D_1 and from D_1 to D_2 seem to be favorable in order to reduce waste entropy. In such cases, decision makers may have a dilemma on what design platform is optimal from the customer's perspective. Therefore, the following decision-making tool to avoid the problem is proposed.

Table 7.2 Computational results of numbers of product configurations

Product design platforms	Number of product configurations	
	Without restrictions n_s (complete design space)	With restrictions n_v (constrained design space)
D_0	108	21
D_1	72	18
D_2	48	18

Table 7.3 Computational results of waste entropy for different numbers of product configurations

	Platform D_0	Platform D_1	Platform D_2
Indicator	Before reduction of components	After withdrawal of MB2	After withdrawal of MB2, CPU486
Entropy of design space	$H_x = \log_2 n_s$ [bit] $n_s = 108$ $H_x = 6{,}75$ bits	$H_x = \log_2 n_s$ [bit] $n_s = 72$ $H_x = 6{,}2$ bits	$H_x = \log_2 n_s$ [bit] $n_s = 48$ $H_x = 5{,}58$ bits
Entropy of constrained design space	$H_c = \log_2 n_v$ [bit] $n_v = 21$ $H_c = 4{,}39$ bits	$H_c = \log_2 n_v$ [bit] $n_v = 18$ $H_c = 4{,}17$ bits	$H_c = \log_2 n_v$ [bit] $n_v = 18$ $H_c = 4{,}17$ bits
Waste entropy	$H_w = H_x – H_c$ $= 2{,}36$ bits	$H_w = H_x – H_c =$ $2{,}03$ bits	$H_w = H_x – H_c = 1{,}4$ bit
Waste entropy ratio	$H_w/H_x = 34{,}9\%$	$H_w/H_x = 32{,}7\%$	$H_w/H_x = 25{,}3\%$

7.3 Decision-making algorithm

In this section, we describe a decision-making procedure to select the optimal platform of product variants by using the mutual relations between waste entropy H_w and constrained design space H_c.

We start by taking the so-called draft design platform D_0, representing an existing product design platform generating both feasible and infeasible product configurations for customers, where n_{s0} is the number of unique product design configurations as a result of a combination of product components and n_{v0} is the number of feasible product design configurations.

Let us further assume that we remove a single component from platform D_0 that is in conflict with other component(s). Then D_0 can be transformed into a new state with n_{s_1} for all unique product design configurations and n_{v_1} for feasible product configurations, denoted as platform D_1.

If such a reduction of components is continued, the design platform D_1 is modified into D_2. Obviously, it is possible to continue the reduction of system components depending on specific conditions.

To compare exactly two arbitrary design platforms with each other (e.g., D_0 and D_1), the following two measures can be proposed.

$$\Delta H_{w0,1} = \left| \frac{H_{w1}}{H_{w0}} - 1 \right| \tag{7.9}$$

$$\Delta H_{c0,1} = \left| \frac{H_{c1}}{H_{c0}} - 1 \right| \tag{7.10}$$

where $\Delta H_{w0,1}$ is the difference between the waste entropies of the design platforms D_1 and D_0, and $\Delta H_{c0,1}$ is the difference between the entropies of the constrained design space of platforms D_1 and D_0.

Then, if $\Delta H_{w0,1} > \Delta H_{c0,1}$, design platform D_1 is more preferable for MC than D_0. To compare three alternative design platforms, the following subprocedure can be used. Let us suppose that design platforms D_1 and D_2 are more preferable for MC than D_0, based on the following criteria:

$$\Delta H_{w0,1} > \Delta H_{c0,1}$$

$$\Delta H_{w0,2} > \Delta H_{c0,1}$$

Then, one can select the more preferable design platform between D_1 and D_2 using the following three criteria:

1. If $\Delta H_{w0,1} - \Delta H_{c0,1} > \Delta H_{w0,2} - \Delta H_{c0,2}$, design platform D_1 is more suitable than D_2.
2. If $\Delta H_{w0,1} - \Delta H_{c0,1} < \Delta H_{w0,2} - \Delta H_{c0,2}$, design platform D_2 is more suitable than D_1.
3. If $\Delta H_{w0,1} - \Delta H_{c0,1} = \Delta H_{w0,2} - \Delta H_{c0,2}$, both design platforms D_1 and D_2 are equally preferable for buyers.

The proposed procedure for selecting an optimal design platform is graphically depicted in Figure 7.5 in the form of a decision-making algorithm.

7.4 Practical case application

In order to prove the relevance of the proposed decision-making tool for selecting the most optimal product design platform, the following realistic case of customized bicycle components is used (Table 7.4). The case application in this section is represented by restrictions between two interoperating modules: the front drivetrain and crankset, which can be found in every bicycle model. Restrictions in these models will be denoted as "X" and represent the incompatibility of the two modules. The starting platform D_0 consists of 12 groups: nine for gears and three for the chain stay angle (CSA). Each of the nine groups has a specific number of alternative components to be combined with the front drivetrain; for example, gear 42-32-24T can be combined with six front cranksets: M980, M780, M670, M610, M552, and M522. With this assignment of components, one could construct a symmetric incidence matrix with size $n = 57$ using the same principle with which the matrix in Section 7.2.2.1 was constructed. In such a case, the symmetrical matrix would be needlessly complex. Therefore, it

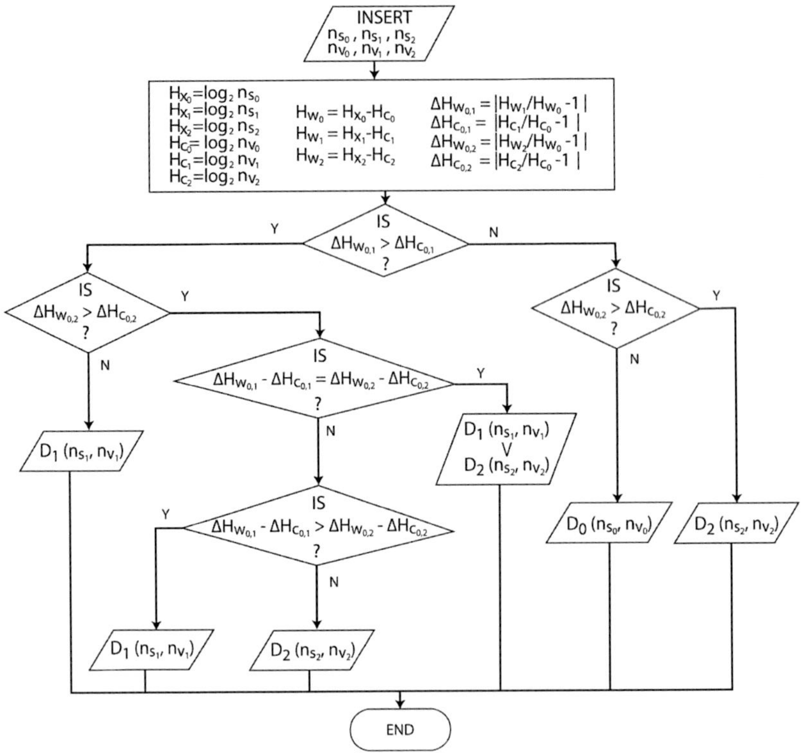

Figure 7.5 Procedure for the selection of an optimal design platform.

is advantageous to keep it in its original form as a nonsymmetrical matrix consisting of 38 rows and 19 columns.

For design platform D_0, the complete design space is determined by $n_{s0} = 722$ product configurations, and the restricted design space is expressed by $n_{v0} = 239$ product configurations.

By using this matrix, it is possible to gradually remove selected component entries with restrictions from product platform D_0 to obtain concurrent platforms.

In order to benchmark all possible concurrent product platforms at once, gears 48-36-26T, including eight cranksets (M610, T780, M670, T781, T671, T611, T551, and T521), have been selected for execution into platform D_1. This group of components was selected for execution based on the criterion of the highest density of restrictions. Subsequently, we obtain a compatibility table, as seen in Appendix 7.1, where gears 48-36-26T are omitted. The number of rows in this table was reduced from 38 to 30.

Obtained platform D_1 consists of 570 drivetrain configurations (n_{s1}) and 215 restricted (viable) product configurations (n_{v1}).

Table 7.4 Module compatibility platform D_0 for gears and front drivetrain

Platform D_0		Front drivetrain																		
		For triple											For double							
CSA		66°–69°								63°–66°			66°–69°							
Gears	Front crankset	M981	M781-A	M671-A	M611	M981-D	M781-A-D	M671-A-D	M611-D	T781-3	T671-3	T611-3	M986	M786	M676	M616	M986-D	M786-D	M676-D	M616-D
42-32-24T	M980									X	X	X	X	X	X	X	X	X	X	X
	M780									X	X	X	X	X	X	X	X	X	X	X
	M670									X	X	X	X	X	X	X	X	X	X	X
	M610									X	X	X	X	X	X	X	X	X	X	X
	M552									X	X	X	X	X	X	X	X	X	X	X
	M522									X	X	X	X	X	X	X	X	X	X	X
40-30-22T	M782	X				X				X	X	X	X	X	X	X	X	X	X	X
	M672	X				X				X	X	X	X	X	X	X	X	X	X	X
	M622	X				X				X	X	X	X	X	X	X	X	X	X	X
	M612	X				X				X	X	X	X	X	X	X	X	X	X	X
	M523	X				X				X	X	X	X	X	X	X	X	X	X	X
48-36-26T	M610	X	X	X	X	X	X	X	X				X	X	X	X	X	X	X	X
	T780	X	X	X	X	X	X	X	X				X	X	X	X	X	X	X	X
	M670	X	X	X	X	X	X	X	X				X	X	X	X	X	X	X	X
	T781	X	X	X	X	X	X	X	X				X	X	X	X	X	X	X	X
	T671	X	X	X	X	X	X	X	X				X	X	X	X	X	X	X	X
	T611	X	X	X	X	X	X	X	X				X	X	X	X	X	X	X	X
	T551	X	X	X	X	X	X	X	X				X	X	X	X	X	X	X	X
	T521	X	X	X	X	X	X	X	X				X	X	X	X	X	X	X	X
44-32-24T	T611	X	X	X	X	X	X	X	X				X	X	X	X	X	X	X	X
	T551	X	X	X	X	X	X	X	X				X	X	X	X	X	X	X	X
	T521	X	X	X	X	X	X	X	X				X	X	X	X	X	X	X	X
44-30T	M985	X	X	X	X	X	X	X	X	X	X	X								
42-30T	M985	X	X	X	X	X	X	X	X	X	X	X								

(*Continued*)

Table 7.4 (Continued) Module compatibility platform D_0 for gears and front drivetrain

Platform D_0		Front drivetrain																		
		For triple											For double							
CSA		66°–69°								63°–66°			66°–69°							
Gears	Front crankset	M981	M781-A	M671-A	M611	M981-D	M781-A-D	M671-A-D	M611-D	T781-3	T671-3	T611-3	M986	M786	M676	M616	M986-D	M786-D	M676-D	M616-D
40-28T	M985	X	X	X	X	X	X	X	X	X	X	X								
	M785	X	X	X	X	X	X	X	X	X	X	X								
	M675	X	X	X	X	X	X	X	X	X	X	X								
	M625	X	X	X	X	X	X	X	X	X	X	X								
	M615	X	X	X	X	X	X	X	X	X	X	X								
38-26T	M980	X	X	X	X	X	X	X	X	X	X	X								
	M785	X	X	X	X	X	X	X	X	X	X	X								
	M675	X	X	X	X	X	X	X	X	X	X	X								
	M625	X	X	X	X	X	X	X	X	X	X	X								
	M615	X	X	X	X	X	X	X	X	X	X	X								
38-24T	M785	X	X	X	X	X	X	X	X	X	X	X								
	M675	X	X	X	X	X	X	X	X	X	X	X								
	M625	X	X	X	X	X	X	X	X	X	X	X								
	M615	X	X	X	X	X	X	X	X	X	X	X								

Afterward, to determine platform D_2, we proceed toward a reduction of the gear type 44-32-24T, including three cranksets (T611, T551, and T521), as can be seen in Appendix 7.2. The number of rows in this table was reduced from 30 to 27.

Then, obtained platform D_2 is determined by $n_{s_2} = 513$ drivetrain configurations and $n_{v_2} = 206$ restricted (viable) product configurations.

In order to provide the next alternative product platform (D_3) for the benchmarking study, two drivetrains, M981 and M981-D, have been eliminated due to the high number of restrictions related to the two components. Then, we obtained platform D_3, defined by $n_{s_3} = 459$ drivetrain configurations and $n_{v_3} = 194$ restricted product configurations (see Appendix 7.3), while the number of columns decreased from 19 to 17.

The obtained numbers of drivetrain configurations with related values of waste entropy H_w and constrained design space H_c (with and without restrictions) are summarily depicted in Table 7.5.

Table 7.5 Computational results of numbers of product configurations

Indicator	Platform D_0	Platform D_1	Platform D_2	Platform D_3
Entropy of design space	$H_x = \log_2 n_s$ [bit] $n_s = 722$ $H_x = 9.50$ bits	$H_x = \log_2 n_s$ [bit] $n_s = 570$ $H_x = 9.15$ bits	$H_x = \log_2 n_s$ [bit] $n_s = 513$ $H_x = 9.00$ bits	$H_x = \log_2 n_s$ [bit] $n_s = 459$ $H_x = 8.84$ bits
Entropy of constrained design space	$H_c = \log_2 n_v$ [bit] $n_v = 239$ $H_c = 7.90$ bits	$H_c = \log_2 n_v$ [bit] $n_v = 215$ $H_c = 7.75$ bits	$H_c = \log_2 n_v$ [bit] $n_v = 206$ $H_c = 7.69$ bits	$H_c = \log_2 n_v$ [bit] $n_v = 194$ $H_c = 7.60$ bits
Waste entropy	$H_w = H_x - H_c$ = 1.59 bits	$H_w = H_x - H_c$ = 1.41 bits	$H_w = H_x - H_c$ = 1.32 bit	$H_w = H_x - H_c$ = 1.24 bit

In the next step, the decision-making algorithm for determining the suitable extent of product variety for different platforms can be applied. Since the algorithm in Figure 7.5 is dedicated to a maximum of three alternative design platforms, an extension of this algorithm for a maximum of four design platforms has been constructed (see Figure 7.6). After using the algorithm, as the most suitable alternative with respect to the amount of waste entropy, platform D_3 is indicated.

7.5 *Concluding remarks*

The proposed novel method can be employed to assist product managers to independently assess competitive product variety platforms against each other and to evaluate their customization characteristics quantitatively. As was shown and proved in multiple cases, the previously described approach based on empirical criteria leads to a decision on the optimal product platform. In other words, a suitable product platform is understood as the structure of product components (compulsory, voluntary, and stable) with minimum waste entropy, and at the same time keeping the so-called positive entropy of the constrained design space (enumerated from feasible options) sufficiently high—that is, options that still satisfy a wide range of requirements.

The development of this method was motivated by previous experiences where restricted options in terms of MC are not perceived positively by individual users. In this context, some authors [6–9] argue that infeasible configurations might be hidden by using algorithm-based product configurators. Quelch and Jocz [10] underline the role of information technology in ensuring that buyers do not choose incompatible options. However, it is evident that one type of configurator engine has been developed especially for options that include infeasible component combinations [11]. Thus, the problem treated in this chapter opens new research

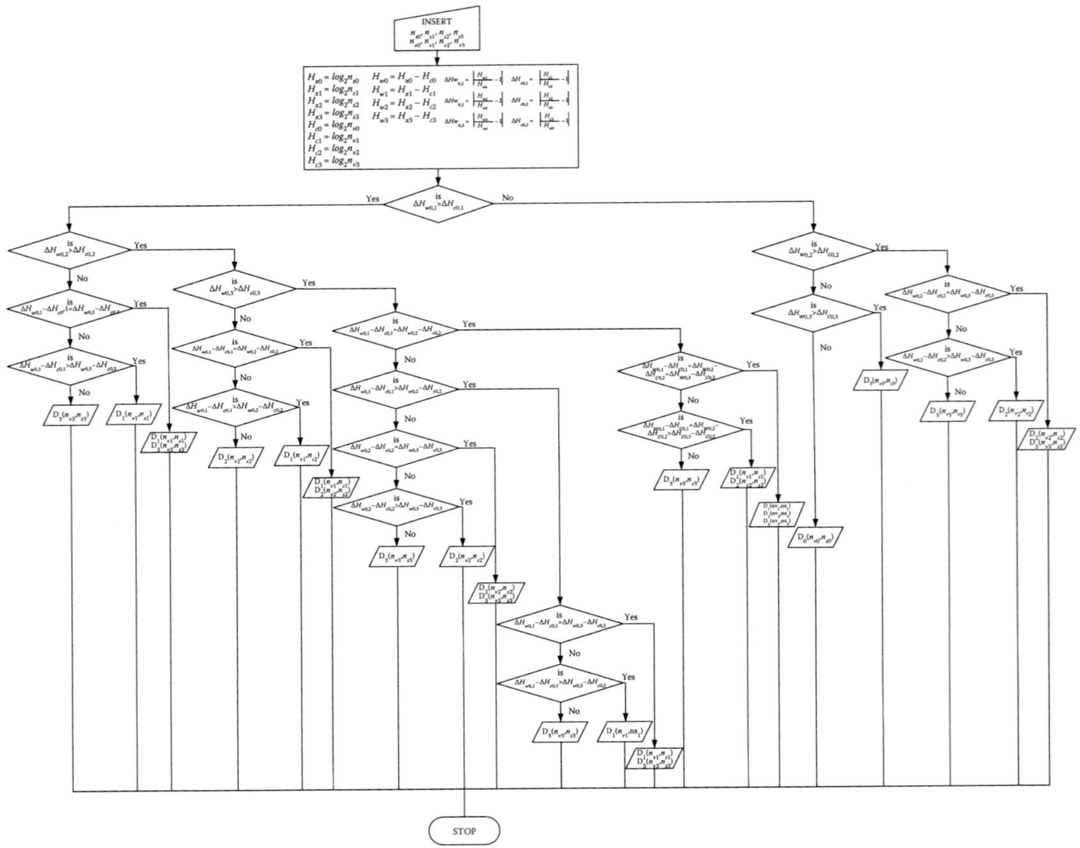

Figure 7.6 Procedure for the selection of an optimal design platform from the three available.

perspectives, as each different sector of MC requires a specific, effective approach to solving configuration conflict problems related to product structures with restricted configuration options.

References

1. Bonev, M., Hvam, L., Clarkson, J., and Maier, A. (2015). Formal computer-aided product family architecture design for mass customization. *Computers in Industry*, 74, 58–70.
2. Shannon, C. E. (1948). A mathematical theory of communication. *Bell System Technical Journal*, 27(3), 379–423.
3. Krus, P. (2015). Design space configuration for minimizing design information entropy. In *ICoRD '15 Research into Design across Boundaries Volume 1*, pp. 51–60. New Delhi, India: Springer.
4. Yang, D., and Dong, M. (2012). A constraint satisfaction approach to resolving product configuration conflicts. *Advanced Engineering Informatics*, 26(3), 592–602.
5. Modrak, V., Krus, P., and Bednar, S. (2015). Approaches to product variety management assuming configuration conflict problem. *FME Transactions*, 43, 271–278.
6. Pitiot, P., Aldanondo, M., and Vareilles, E. (2014). Concurrent product configuration and process planning: Some optimization experimental results. *Computers in Industry*, 65(4), 610–621.
7. Helo, P. T., Xu, Q. L., Kyllonen, S. J., and Jiao, R. J. (2010). Integrated vehicle configuration system connecting the domains of mass customization. *Computers in Industry*, 61(1), 44–52.
8. Mailharro, D. (1998). A classification and constraint-based framework for configuration. *Artificial Intelligence for Engineering Design, Analysis and Manufacturing*, 12(4), 383–397.
9. Aldanondo, M., Vareilles, E., and Djefel, M. (2010). Towards an association of product configuration with production planning. *International Journal of Mass Customisation*, 3(4), 316–332.
10. Quelch, J. A., and Jocz, K. E. (2013). *Greater Good: How Good Marketing Makes for Better Democracy*. Boston, MA: Harvard Business Press.
11. Orsvärn, K., and Axling, T. (1999, April). The Tacton view of configuration tasks and engines. In *Workshop on Configuration, Sixteenth National Conference on Artificial Intelligence (AAAI-99)*. Palo Alto, CA: Association for the Advancement of Artificial Intelligence.

Appendix 7.1

Reduced module compatibility platform D_1

Platform D_1		Front drivetrain																		
		For triple											For double							
CSA		66°–69°								63°–66°			66°–69°							
Gears	Front crankset	M981	M781-A	M671-A	M611	M981-D	M781-A-D	M671-A-D	M611-D	T781-3	T671-3	T611-3	M986	M786	M676	M616	M986-D	M786-D	M676-D	M616-D
42-32-24T	M980									X	X	X	X	X	X	X	X	X	X	X
	M780									X	X	X	X	X	X	X	X	X	X	X
	M670									X	X	X	X	X	X	X	X	X	X	X
	M610									X	X	X	X	X	X	X	X	X	X	X
	M552									X	X	X	X	X	X	X	X	X	X	X
	M522									X	X	X	X	X	X	X	X	X	X	X
40-30-22T	M782	X				X				X	X	X	X	X	X	X	X	X	X	X
	M672	X				X				X	X	X	X	X	X	X	X	X	X	X
	M622	X				X				X	X	X	X	X	X	X	X	X	X	X
	M612	X				X				X	X	X	X	X	X	X	X	X	X	X
	M523	X				X				X	X	X	X	X	X	X	X	X	X	X
44-32-24T	T611	X	X	X	X	X	X	X	X				X	X	X	X	X	X	X	X
	T551	X	X	X	X	X	X	X	X				X	X	X	X	X	X	X	X
	T521	X	X	X	X	X	X	X	X				X	X	X	X	X	X	X	X
44-30T	M985	X	X	X	X	X	X	X	X	X	X	X								
42-30T	M985	X	X	X	X	X	X	X	X	X	X	X								
40-28T	M985	X	X	X	X	X	X	X	X	X	X	X								
	M785	X	X	X	X	X	X	X	X	X	X	X								
	M675	X	X	X	X	X	X	X	X	X	X	X								
	M625	X	X	X	X	X	X	X	X	X	X	X								
	M615	X	X	X	X	X	X	X	X	X	X	X								
38-26T	M980	X	X	X	X	X	X	X	X	X	X	X								
	M785	X	X	X	X	X	X	X	X	X	X	X								
	M675	X	X	X	X	X	X	X	X	X	X	X								
	M625	X	X	X	X	X	X	X	X	X	X	X								
	M615	X	X	X	X	X	X	X	X	X	X	X								
38-24T	M785	X	X	X	X	X	X	X	X	X	X	X								
	M675	X	X	X	X	X	X	X	X	X	X	X								
	M625	X	X	X	X	X	X	X	X	X	X	X								
	M615	X	X	X	X	X	X	X	X	X	X	X								

Appendix 7.2

Reduced module compatibility platform D_2

Platform D_1		Front drivetrain																		
		For triple											For double							
CSA		66°–69°								63°–66°			66°–69°							
Gears	Front crankset	M981	M781-A	M671-A	M611	M981-D	M781-A-D	M671-A-D	M611-D	T781-3	T671-3	T611-3	M986	M786	M676	M616	M986-D	M786-D	M676-D	M616-D
42-32-24T	M980									X	X	X	X	X	X	X	X	X	X	X
	M780									X	X	X	X	X	X	X	X	X	X	X
	M670									X	X	X	X	X	X	X	X	X	X	X
	M610									X	X	X	X	X	X	X	X	X	X	X
	M552									X	X	X	X	X	X	X	X	X	X	X
	M522									X	X	X	X	X	X	X	X	X	X	X
40-30-22T	M782	X				X				X	X	X	X	X	X	X	X	X	X	X
	M672	X				X				X	X	X	X	X	X	X	X	X	X	X
	M622	X				X				X	X	X	X	X	X	X	X	X	X	X
	M612	X				X				X	X	X	X	X	X	X	X	X	X	X
	M523	X				X				X	X	X	X	X	X	X	X	X	X	X
44-30T	M985	X	X	X	X	X	X	X	X	X	X	X								
42-30T	M985	X	X	X	X	X	X	X	X	X	X	X								
40-28T	M985	X	X	X	X	X	X	X	X	X	X	X								
	M785	X	X	X	X	X	X	X	X	X	X	X								
	M675	X	X	X	X	X	X	X	X	X	X	X								
	M625	X	X	X	X	X	X	X	X	X	X	X								
	M615	X	X	X	X	X	X	X	X	X	X	X								
38-26T	M980	X	X	X	X	X	X	X	X	X	X	X								
	M785	X	X	X	X	X	X	X	X	X	X	X								
	M675	X	X	X	X	X	X	X	X	X	X	X								
	M625	X	X	X	X	X	X	X	X	X	X	X								
	M615	X	X	X	X	X	X	X	X	X	X	X								
38-24T	M785	X	X	X	X	X	X	X	X	X	X	X								
	M675	X	X	X	X	X	X	X	X	X	X	X								
	M625	X	X	X	X	X	X	X	X	X	X	X								
	M615	X	X	X	X	X	X	X	X	X	X	X								

Appendix 7.3

Reduced module compatibility platform D_3

Platform D_3		Front drivetrain																
		For triple																
CSA		66°–69°						63°–66°			66°–69°							
Gears	Front crankset	M781-A	M671-A	M611	M781-A-D	M671-A-D	M611-D	T781-3	T671-3	T611-3	M986	M786	M676	M616	M986-D	M786-D	M676-D	M616-D
42-32-24T	M980							X	X	X	X	X	X	X	X	X	X	X
	M780							X	X	X	X	X	X	X	X	X	X	X
	M670							X	X	X	X	X	X	X	X	X	X	X
	M610							X	X	X	X	X	X	X	X	X	X	X
	M552							X	X	X	X	X	X	X	X	X	X	X
	M522							X	X	X	X	X	X	X	X	X	X	X
40-30-22T	M782							X	X	X	X	X	X	X	X	X	X	X
	M672							X	X	X	X	X	X	X	X	X	X	X
	M622							X	X	X	X	X	X	X	X	X	X	X
	M612							X	X	X	X	X	X	X	X	X	X	X
	M523							X	X	X	X	X	X	X	X	X	X	X
44-30T	M985	X	X	X	X	X	X	X	X	X								
42-30T	M985	X	X	X	X	X	X	X	X	X								
40-28T	M985	X	X	X	X	X	X	X	X	X								
	M785	X	X	X	X	X	X	X	X	X								
	M675	X	X	X	X	X	X	X	X	X								
	M625	X	X	X	X	X	X	X	X	X								
	M615	X	X	X	X	X	X	X	X	X								
38-26T	M980	X	X	X	X	X	X	X	X	X								
	M785	X	X	X	X	X	X	X	X	X								
	M675	X	X	X	X	X	X	X	X	X								
	M625	X	X	X	X	X	X	X	X	X								
	M615	X	X	X	X	X	X	X	X	X								
38-24T	M785	X	X	X	X	X	X	X	X	X								
	M675	X	X	X	X	X	X	X	X	X								
	M625	X	X	X	X	X	X	X	X	X								
	M615	X	X	X	X	X	X	X	X	X								

section three

Management and sustainability of mass customization

chapter eight

Product configuration for order acquisition and fulfillment

Dario Antonelli and Giulia Bruno

Contents

ABSTRACT

A key issue for the effective application of *mass customization* (MC) design principles is associated with the order acquisition and fulfillment process, which becomes increasingly more complex with the proliferation of distinct product variants. Nowadays, the process cannot do without the assistance of a *product configurator* system. The evolution of the product configurator toward a tool able to support the design of new product variants is a challenging problem, not fully solved in a large variety of production sectors.

An underestimated issue in the implementation of a product configurator derives from the lack of vision encompassing the whole *product life cycle* during the early phases of product definition. As a matter of fact, the design of a new product variant should not omit consideration of the technical and economic feasibility of product manufacturing and its impact on the *capacity planning* and *material resource planning* of the whole industrial plant.

This chapter describes the problem by means of interactions with a metamodel of a generic product family. The metamodel has the form of an ontology and will enable interactive software (namely, the product configurator) to assist the customer during the definition of a customized product. The metamodel should give the rules to build every possible product tree (a reproduction of the bill of material [BOM]) for every product variant in the family.

The *product life cycle management* (PLM) approach extends the product description to the process constraints and production requirements in order to simultaneously solve product configuration and order fulfillment problems.

8.1 Introduction

Configurators are software applications intended to help both the customer and the dealer at the stage of order acquisition and fulfillment (Forza and Salvador, 2002). They are largely applied in industrial sectors, such as computer, automotive, and clothing companies. Their objective is to help the customer to configure a personalized version of the product and conversely to ensure a fast, complete, and reliable transmission of the order to production. As a matter of fact, the growth of customizable features of products has forced a far larger growth in the amount and variety of product data that has to be communicated from the customer to the company before the execution of the order. Therefore, it is helpful to use the support of an automatic system dealing with the number and variety of product features and their related data. The product configurator translates the customer choices into a coherent BOM. It controls the availability of every optional component of the product and detects conflicting components—that is, components that cannot be used/assembled together in the final product.

Configurable products can be defined as predesigned products that (Tiihonen et al., 1996)

- Are adapted according to the requirements of the customer for each order
- Consist of (almost) only predesigned components
- Have a predesigned product structure
- Are adapted by a routine, systematic product configuration process

Each configurable product has to be described by a product configuration model that encompasses the product trees of every individual

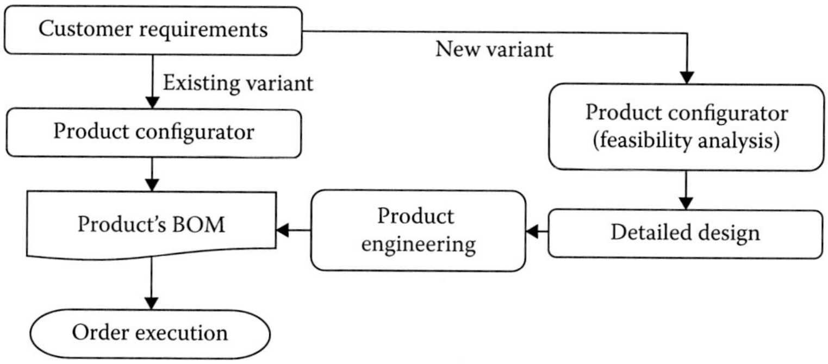

Figure 8.1 Comparison of the two approaches to product configuration.

product. The computer system supporting the product configuration tasks, the product configurator, must be able to utilize product configuration models.

A main distinction can be made between the application of the product configurator to help the customer in selecting among several product alternatives already existing in the company catalog and its application as a support tool to design and develop a new variant on the basis of the customer's specifications (Figure 8.1). The second alternative, which can be called the *generative* approach, is far more challenging and difficult to implement in industrial sectors characterized by continuous innovation or high product complexity. Nevertheless, it would be particularly useful in assisting the order acquisition of a large variety of industrial products that have to be tailored to customer specifics, such as commercial vehicles, data servers, clothes, furniture (kitchens, libraries), or machine tools.

In the generative case, the product configurator has two tasks: firstly, to support the customer in the setup of a feasible product variant by enforcing the application of the design constraints; secondly, to fulfill the order by assembling all the parts into a new product variant in order to generate the corresponding BOM.

There are two approaches to the design of a product family: top down and bottom up (Farrel and Simpson, 2003). In the top-down approach, the company develops single products after deciding on a set of common features that are used in all the family products. In the bottom-up approach, different products are redesigned in order to include them in a family, increasing the number of common features among them. The first approach gets the maximum advantage from the product platform (Muffatto, 1999; Meyer and Dhaval, 2002).

To achieve the scope of a fully generative product configurator delivering reliable output without significant human intervention, there are two competing approaches: increase the intelligence of the configuration software or simplify the complexity of the product by modularity.

The first technique refers to writing an expert system made of a knowledge base containing all the rules necessary to correctly assemble different items in order to generate a new product from a library of subassemblies. An inferential engine would query the knowledge base to comply with the customer requests and to check their feasibility. The first expert system, XCON, introduced by Digital Research, became a typical example of the successful application of expert systems technology. XCON was a good illustration of the complexity of knowledge maintenance. In 1989, its knowledge base had more than 31,000 components and approximately 17,500 rules (Sabin and Rainer, 1998).

The second technique refers to the simplification of the assembly interfaces to a point that will not require any *intelligence* to assemble the different items and where the procedure will be implemented on existing configurator software. To do this, it is necessary to redesign all the existing families of products and to accept the inevitable shortcomings of a more expensive and less optimized product.

Nevertheless, there is a third way that has gained popularity since the beginning of the century: to embed intelligence inside the product description. If the product tree contains not only the hierarchy of the assembly but also the constraints to be followed and the producing techniques, it is possible to analyze a new product request and to generate the corresponding specifications of the process plan by smart application of the production rules embedded in the product tree. This approach requires extra effort of the product designer because he or she has not only to design a number of parts belonging to the product family, but also to make explicit the constraints and incompatibilities that link each of the various items that compose a product. The advantage is that it does not require a redesign of the existing products, and it is possible to adapt the existing product configurator to a more structured set of building rules.

This approach has seen a consistent evolution. It started from the building of product metatrees, which are a set of rules to build a tree of all the products belonging to the same family (Olsen et al., 1997). Then, it was proposed to adopt an object-oriented representation of the product in order to exploit the advantages of object-oriented philosophy: inheritance, easily reusable models, and functional descriptions of the product components (Hedin et al., 1998). This approach naturally led to a representation of the product tree in terms of the Unified Modeling Language (UML; http://www.uml.org) class diagram or, more generally, by modeling the classes and data types in a representation independent of their implementation. The final arrival point was the representation of the ontology of the product (Stark, 2005).

In order to clarify the rationale behind this evolving pattern, it is useful to take a step backward and describe in detail the meaning of BOM and its constitutive rules.

8.2 Product tree and bill of materials

Configuring a product implies the selection of items from a set of subcomponents and their organization and structure in order to build a functional assembly. The final product has to satisfy a number of requirements that derive from the customer, from the enterprise's dealer, or from the conceptual definition of a new product. Put simply, configuration is an activity whose input is determined by the customer requests specifying the peculiar variant of the product he or she is aiming to buy. The output is the generation of a document containing a description of the product, mapping the specifications of the customer. This document is the BOM. A BOM is organized hierarchically, with a root indicating the product and a number of branches representing the subassemblies that compose the product. From every branch, other sub-branches can depart, allowing for many different detail levels. The description is made in terms of assembly, subassembly, or individual items. A BOM is graphically represented as a product tree: a graph describing ordered networks of interconnected nodes. Visually, it resembles an inverted tree.

For the sake of discussion, let us consider the example of Figure 8.2, which shows the product tree of a personal computer.

If all of the nodes in the graph are sorted from top to bottom, starting from the end node (which corresponds to the assembled product), it derives a representation of the product in terms of its components. For each component at a given level, it is possible to associate simpler derived components at a lower level. Level 1 refers to the final product; level 2 refers to the subassembled components, and so on downward, until reaching a part bought as is from external providers or produced internally from raw material. The BOM corresponding to the product tree of Figure 8.2 is reported in Table 8.1.

The extension of the product description to the product family description requires a generalization step that can be named the *product metatree* (Antonelli and Villa, 2006).

Let us consider a family of products with a still-undefined number of components. Each product configuration is represented by the BOM. Every item inside the BOM is an object defined by a number of geometric and functional attributes (states) and by the techniques used to produce/acquire/assemble it (behaviors). The representation of the product family requires the design of a superclass that encloses all items of the family. The superclass is equivalent to a product metatree—that is, a product tree with only the branch structure and the levels but without instancing the items corresponding to one specific product. The connections among different items on different levels are described by means of logic operators: AND, OR, and XOR. It is called a metatree because it represents all the possible product trees of a product family by giving the connection

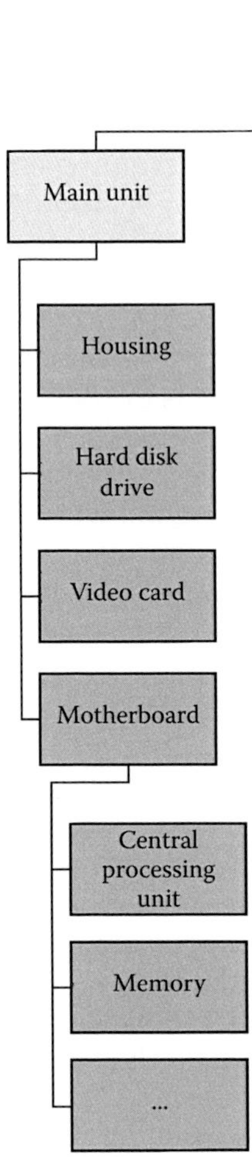

Figure 8.2 Product tree of a computer.

rules. In Figure 8.3, a sample family tree is represented together with the corresponding functional tree, which expresses the functions assigned to every component.

So far, a single product has not been defined but a set of functional requisites matched with corresponding physical components. The pair of metatrees gives a thorough description of a family of products in terms of the functionalities that can be performed by different product instances and the contribution of every component to the product operations, as well as in terms of the assembly configuration of each product. A product family is described in terms of a function vector—that is, the list of

Table 8.1 BOM corresponding to the product tree of Figure 8.2

Bill of materials: Product computer			
Level	Code	Description	Quantity
1	1-0-0	Computer	1
2	1-1-0	Main unit	1
2	1-2-0	Monitor, LED, 28", full HD	1
2	1-3-0	Mouse, wireless optical	1
2	1-4-0	Keyboard, wireless	1
3	1-1-1	Housing, ATX	1
3	1-1-2	Hard disk, SATA, 1 TB	2
3	1-1-3	Video card, 2 GB, GDDR5	1
3	1-1-4	Motherboard	1
4	1-1-4-1	Motherboard plate, ATX	1
4	1-1-4-2	CPU, 64 bit, 4 GHz	1
4	1-1-4-3	RAM card, 4 GB	2
4	…	…	…

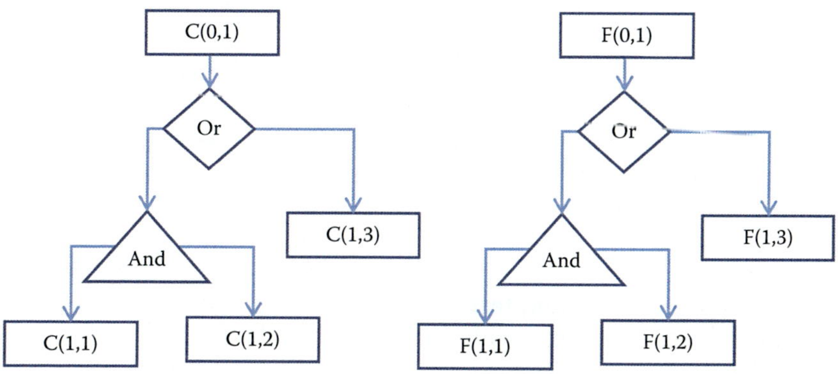

Figure 8.3 Examples of a component tree and a function tree.

functions related, respectively, to the product family F(0,1) and to each instance j on a given detail level i: F(i,j).

Therefore, the function tree can be modeled by a connection matrix—that is, an incidence matrix that specifies both the existence of a connection between two nodes in the tree and the type of connection. The connection matrix MF is generated by the application of the following rules:

- MF{F(i,j),F(n,m)} = 1 if F(i,j) is a subitem of F(n,m).
- MF{F(i,j),F(n,m)} = AND if F(i,j) has to be joined with F(n,m).
- MF{F(i,j),F(n,m)} = OR if F(i,j) is an alternative to F(n,m).
- MF{F(i,j),F(n,m)} = XOR if F(i,j) is the exclusive alternative to F(n,m).

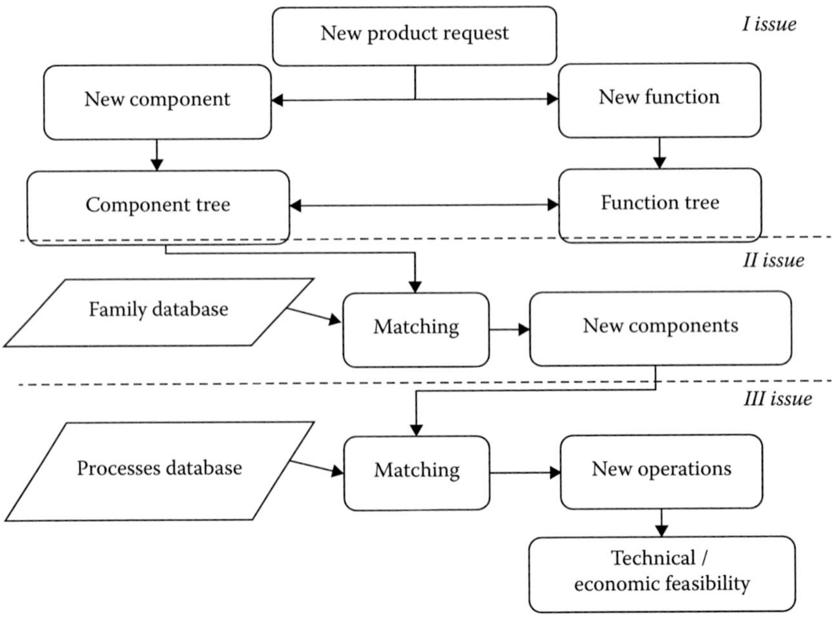

Figure 8.4 Decomposition of the problem in three main issues.

The connection matrix is asymmetric and is composed of a lower triangular matrix with the product structure and an upper triangular matrix with the connections.

The next step is creating an object-oriented representation of the superclass of products by assigning to every component one or more functions and its geometric and technical attributes (Figure 8.4). This is done in STEP, the standard representation of product data using the EXPRESS formal language.

Now, let us assume that the customer proposes a new variant in terms of new function requirements and/or new components, selected in the component library. The configuration of a new product variant is composed of three subsequent issues. (Figure 8.4).

The first issue is the making of a product and function tree corresponding to the desired product variant. This is equivalent to instancing the class of product trees corresponding to the product family. The solution of the first problem is obtained by a thorough application of object-oriented methods: mapping the functions of the new project onto the functions of the product family, masking class functions not required by the customer, and checking the compatibility of the corresponding interfaces. The first step is made possible by the rule of inheritance inside an object class. As a matter of fact, if the customer asks for a new function that substitutes an existing one, all the functions that are in a lower hierarchical level will be inherited by the new function.

The second issue is to draft a process-planning method in order to verify the feasibility. This should be obtained by the transformation of the product tree in a corresponding manufacturing sequence. The structure of the product tree is directly converted into a manufacturing tree by applying the concept of *pattern recognition.*

The third issue is to define the schedule of the production process of the new product variant in order to estimate its production lead time and the manufacturing cost. For every operation, the production lead times and costs are given; therefore, it is possible to define the completion time of the entire product and the productive cost.

As an example, let us consider a case study: an electric mixer to be employed in the preparation of different kinds of food. The high-level component tree is provided in Figure 8.5, where a circle substitutes the AND, a rhombus the OR, and functions are in italics. The mixer is obtained by assembling together the individual items, with a choice between two socket variants: "Thread" for a slow but secure connection and "Connection" for a fast connect/disconnect operation. It is worth remembering that functions are inherited by objects along the relationship chain. As an example, the cover inherits the supporting function from its parent, the base of the mixer. Let us now suppose that the customer needs a universal adapter for the same mixer. It is a common component, available in the company database, but it has not been provided in this case. A conventional product configurator would discard the request, while the generative one can generate a new product tree for this variant, replacing the discarded function. The component will inherit all the functions owned by the parent items; therefore, its individual connection matrix will be mapped onto the family connection matrix of Table 8.2 (the difference being in fact the substitution of AND and OR with 1 or 0).

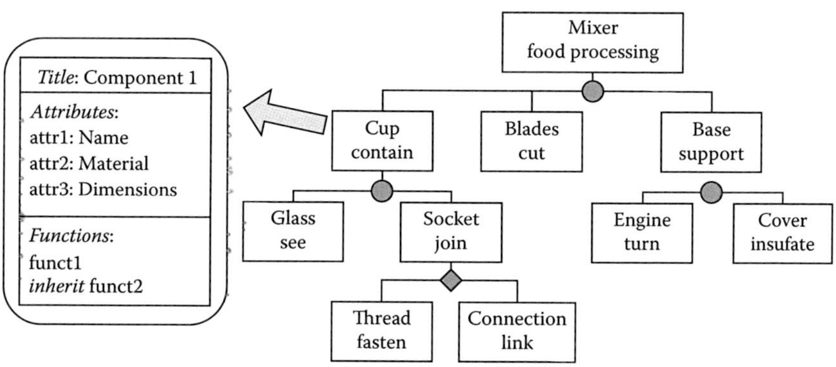

Figure 8.5 Class diagram of a component and its application to the mixer example.

Table 8.2 Individual connection matrix for the mixer with the universal adapter

From/To	Proc.	Cont.	Cut	Sup.	See	Join	Work	Ins.	Fast	Link
Process										
Contain	1		A	A						
Cut	1			A						
Support	1									
See		1				A				
Join		1								
Work				1				A		
Insulate				1						
Fasten										
Standard link						1				

8.3 Approaches to model product configurators

There are many intersecting study fields involved in the conception of a product configurator. On the side of product design, the guidelines of mass customization have to be followed: modularity (Hsuan, 1999), commonality (Agard and Tollenaere, 2002), and postponement (Markham et al., 2001). All of these concepts are known and overall accepted and have been employed together with object-oriented design, for example, in the studies of Grady and Liang (1998), Huang and Kusiak (1998), and Jiao et al. (2003).

8.3.1 Mass customization

The modular conception aims at designing product elements with functional overcapabilities (i.e., they are able to satisfy a redundant set of requirements) in such a way as to be interchangeable: modules. The commonality is the generalized use of the same module inside many different products. Modularity depends on two product characteristics: similarity among physical and functional architectures and the minimization of interaction among the components. Some researchers, such as He et al. (1998) or Jiao and Tseng (1999), propose methods aimed at building modular product families. In particular, He et al. propose a matrix decomposition of the product tree in order to highlight the interchangeable or independent elements. Jiao and Tseng propose a method to develop the architecture of product families by applying three different points of view—functional, technical, and physical—to the classification of product variants.

8.3.2 Object-oriented design

In the object-oriented approach, an object is a bundle of related variables and methods (Eden, 2002). Objects are a model of the real-world objects, with a state and a behavior. A class is a blueprint or prototype that defines the variables and the methods common to all objects of a certain kind. Every object is obtained by instancing a class. Classes can be defined in terms of other classes. Each lower-level class (subclass) inherits its state from the upper-level class (superclass).

8.3.3 Unified Modeling Language class diagram

Class structures are graphically represented using the UML class diagram (Fowler and Scott 2000). UML is widely adopted as a specification and modeling formalism in various fields. It evolved as a strong convergence among different competing but semantically very similar specification languages and formalisms that were independently developed by software engineers. It is now standardized by the Object Management Group (OMG). Currently, UML is a *de jure* and *de facto* standard for describing software artifacts, from the initial conception to the specification, design, implementation, testing, and deployment phases. UML is based on a set of several graphical diagram types (Figure 8.6), each with its specific semantics, that allow the expression of static or dynamic system behavior. The graphical notation is extensible through the stereotyping mechanism (i.e., new semantics are defined with an existing symbol by attaching a <<label>> to it), and a corresponding textual representation, for information interchange between different tools and automatic information processing by user-defined procedures, is also standardized by the OMG (XML Metadata Interchange [XMI]). Finally, a formally specified constraint language is defined for UML models, which help us to express domain-specific constraints and complex interdependencies.

The evolution of UML as a specification formalism specifically oriented to software artifacts is evident from the strong support of object-oriented constructs (classes, properties, methods, inheritance, interfaces, etc.), particularly suited to the high degree of flexibility possessed by software systems. However, thanks to the completeness of the formalism, its widespread popularity, and several tools supporting it, UML has also been successfully applied to nonsoftware domains. For example, the automotive industry developed Automotive UML (von der Beeck et al., 2003) to model the complex interaction of active subsystems in a modern vehicle. Also, real-time systems can be modeled thanks to an extension called Embedded UML (Martin et al., 2001), and manufacturing systems

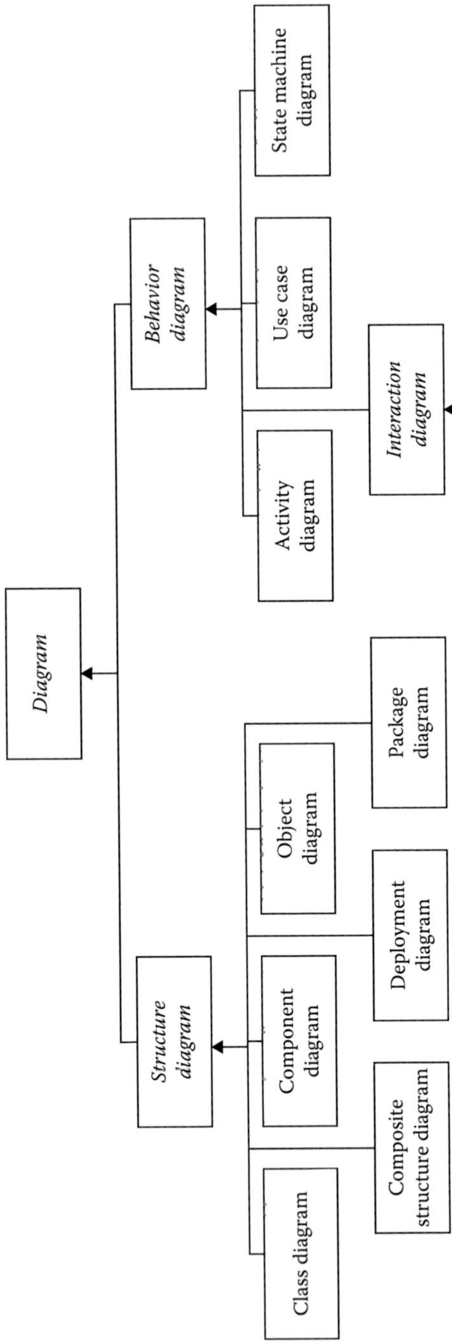

Figure 8.6 Set of UML 2.1 standard diagrams.

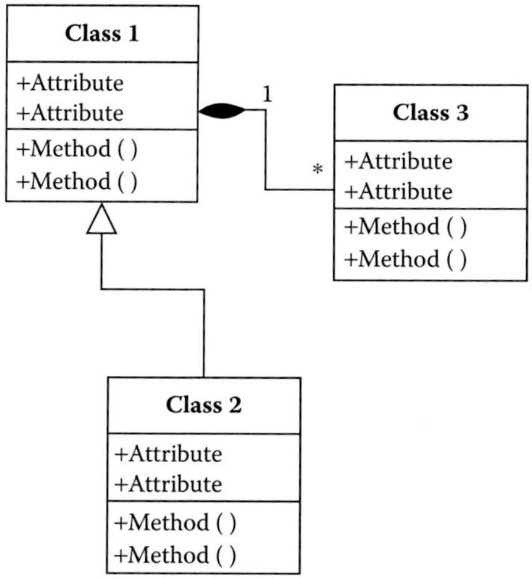

Figure 8.7 An example of class diagram notation.

can be described and simulated with UML representations (Barbarisi et al., 2004).

Figure 8.7 shows an example of a UML class diagram. Each class is drawn as a rectangle with three sections inside. The name section at the top holds the object name. The attribute section in the middle holds a list of the component's features. The operation section at the bottom holds a list of processes to be applied to that object. Classes can be connected through different kinds of relationships (generic associations, aggregation, composition, generalization, etc.), represented as lines with different kinds of end point terminators (arrows, diamonds, etc.).

On the side of the product description methods, the information conveyed through the standard BOM is not sufficient to drive the generation of new product alternatives. Product components can be accurately described (Demartini, 1998) by using the three-layer PDES/STEP methodology (ISO 10303), including the reference model, the format object class and schema definition language EXPRESS, and the file structure. Incidentally, it can be noted that EXPRESS follows an object-oriented paradigm. The adoption of UML diagrams inside the STEP standard is also being discussed by the international standards organization ISO/TC 184/SC 4. While the STEP standard is the preferable choice for the communication of a detailed product description, its detail level is nonetheless too high to allow the efficient implementation of a generative product configurator. Therefore, the solution is to somewhat force the structure of STEP inside a low-detail description of the product.

8.3.4 Ontology representations

As described by Ardito et al. (2011), the idea of exploiting ontology-based product configurators has the advantage of giving customers more freedom in creating products that best fit their desires. The authors analyzed a case study of a company producing pieces of furniture directly ordered by customers, who look at the company catalogs and provide a sketch of each piece of furniture they want, which may be composed of parts chosen from different items in the catalogs and assembled together. In this way, customers have much more freedom in designing their furniture, thanks to the use of an ontology that models the possible composition of different parts in a whole piece.

The ontology was used for two scopes: (1) to connect the objects from different suppliers' catalogs, so that customers can consider all of them as a unique catalog, and (2) to describe the components of each piece of furniture and their properties (e.g., colors, size, decorations, shapes, and materials). The integration of heterogeneous information sources implies the design of a data integration system aimed at dealing with data residing in several sources and at hiding to the user the source of the data he or she is accessing and its structure. Moreover, the ontology provides the rules and constraints to be applied to assemble various components in order to generate only those pieces of furniture that are considered by the ontology.

8.3.5 Proposed approach

The proposed work extends the previous approaches that are only focused on the early stages of the product life cycle (i.e., product design) by proposing an ontology covering the whole product life cycle, in order to assess the influence of the choices made during product configuration on the later stages of the product life cycle—that is, manufacturing, maintenance, and recycling.

8.4 Product life cycle

Product life cycle is the term used to describe the stages that a product goes through in its life. The life cycle is an intrinsic attribute of a physical product.

This section defines the concept of a product life cycle, provides a summary of the ontology that has been developed in this domain, and defines a reference ontology to be used as a common model for the basis of a product configurator.

8.4.1 Product life cycle definition

Although there is a generally accepted product life cycle definition, the product life cycle is still not a very clear or exact concept, because it depends on the perspective from which it is analyzed. For example, from a marketing perspective, the product life cycle covers the phases of introduction, growth, maturity, and decline (Anderson and Zeithaml, 1984). In this context, the evolution of the product is represented from the manufacturer's perspective (Stark, 2005); thus, the following main phases are considered.

- *Design*: The first stage of the product life cycle is the definition of its requirements based on customer, company, market, and regulatory bodies' viewpoints. From this specification of the product, major technical parameters can be defined. This phase also involves functional analysis, requirements allocation, the definition of product components and assembly processes, and development specification.
- *Manufacturing*: Manufacturing (or production) is the physical construction of the product and product components, including acceptance testing, operational testing, and evaluation assessment. Once the design of the product's components is complete, the method of manufacturing is defined. This includes computer-aided design (CAD) tasks such as tool design, the creation of computer numeric control (CNC) machining instructions for the product's parts, as well as tools to manufacture those parts, using integrated or separate computer-aided manufacturing (CAM) software. Once the manufacturing method has been identified, computer-aided process planning (CAPP) comes into play. This involves computer-aided production engineering or production-planning tools for carrying out factory, plant, and facility layout and production simulation. Once components are manufactured, their geometrical form and size can be checked against the original CAD data with the use of computer-aided inspection equipment and software.
- *Maintenance*: Maintenance regards the product operation in the user environment. It includes the management of service information—that is, providing customers and service engineers with support information for repair and maintenance. This involves using tools such as maintenance, repair, and operations (MRO) management software. It also includes performing routine actions that keep the device in working order or prevent trouble from arising.
- *Recycling*: Recycling is the process of converting old products (waste) into new products to prevent the waste of potentially useful materials, reduce the consumption of fresh raw materials, reduce energy usage, and reduce pollution.

8.4.2 Ontologies for product knowledge management

Ontologies have already been proposed for knowledge management in a number of papers (e.g., Jurisica et al., 2004). Since PLM is a complex organization of activities for product and process engineering management, starting from conceptual design, detailed design, engineering, production, and usage, including disposal and recycling, the main reasons for the development of ontology models for PLM are the need for a clear understanding of the product life cycle phases (McKenzie-Veal et al., 2010) and the need for systems interoperability (Abdul-Ghafour, 2012). These reasons motivate the ontology-based data and knowledge representations to support collaborations among the actors operating along the product life cycle, resulting in expected lower efforts and a shorter time to market.

In order to support system integration, approaches dedicated to studying and developing semantic data models with concepts, relations, and their respective properties have been introduced (Fiorentini et al., 2007; Kwak and Yong, 2008). Ontologies have also been developed for STEP (Wang et al., 2010). A product design ontology that formalizes the functionality of shape-processing methods in the design workflow is defined by Catalano et al. (2009), while an ontology to manage both the product and the PLM is proposed by Matsokis and Kiritsis (2010) to automatically handle data from multiple physical products and to support system interoperability and data integration.

The application of ontologies in product development is also fundamental to knowledge sharing (Lutters et al., 2001; Imran and Young, 2013). Lutters et al. (2001) developed an information management system based on an ontological approach to the design and engineering processes, while Imran and Young (2013) showed the benefits of applying ontologies to support knowledge sharing in PLM with a focus on manufacturing processes. By using a product ontology, Panetto et al. (2012) introduced an approach to support interoperability in product data management.

In Sim and Duffy (2003), an ontology for engineering design activities was described, even though it was not implemented in software. An approach to conflict mitigation in collaborative engineering using ontologies implemented in a software module was presented by Slimani et al. (2006). Another approach, also supported by software, was developed by Lu et al. (2008) to allow the management of engineering drawings in the context of shipbuilding. The work presented by Eckstein and Josefiak (2008) describes the SWOP project, which focused on the application of ontologies and problem-solving methods to develop optimized combinations for products and production systems in order to solve customer-specific problems.

8.4.3 Reference ontology for product life cycles

According to Imran and Young (2013), the process to define an ontology starts from the identification of a set of relevant concepts, then proceeds with the organization of concepts in a formal model representing the ontology structure, and finally performs the implementation of the model in ontology web language (OWL). This procedure was followed to define the reference product life cycle ontology.

The set of high-level concepts needed to represent product life cycle knowledge are the following (see Antonelli et al. [2014] and Bruno et al. [2014] for more details):

- *Production item*: Either a product or a product component. Each product can be made of several components, and the same component can be used by different products. Each component can in turn be composed of other components.
- *Characteristic*: A material, a functional characteristic, or a physical characteristic (e.g., height, length, width, and weight) that refers to a production item.
- *Operation*: A manufacturing operation executed by one or more resources to produce a product.
- *Resource*: An entity that is involved in the execution of an activity. It can be of two kinds, person or machine.
- *Role*: The role of a person denoting his or her skills in the company.
- *Failure*: A failure that occurred with a product.
- *Recycling procedure*: The procedure to follow to recycle a component.

These concepts were organized into a formal model, shown in Figure 8.8, according to the UML class diagram formalism (Fowler and Scott, 2000). The rectangles identify the concepts involved in each life cycle phase.

The *production item* class is linked to the *customer* class to store the customer who ordered each product. A production item is also associated with the *characteristic* class to store the characteristic of the item. A production item can be a product or a component. Each production item can be composed of several components.

A product is associated with the operations needed to produce it. To keep track of the resources involved in the operations, the *resource* class is linked to the *operation* class. Also, the other specialization of the resource class (i.e., the *machine* class) is linked to the operation class to store the machines used in each activity. For each person, the *role* he or she has in the company is known, and the roles that can execute each activity are also stored.

For each product, a list of *failures* reported by users is recorded, and for each component the *recycling procedure* is known, if available.

Figure 8.9 shows the implementation of the reference ontology in Protégé.

Once the model to represent a general product life cycle is designed, it can be further specified depending on the industrial domain of application. This process is explained in the following section.

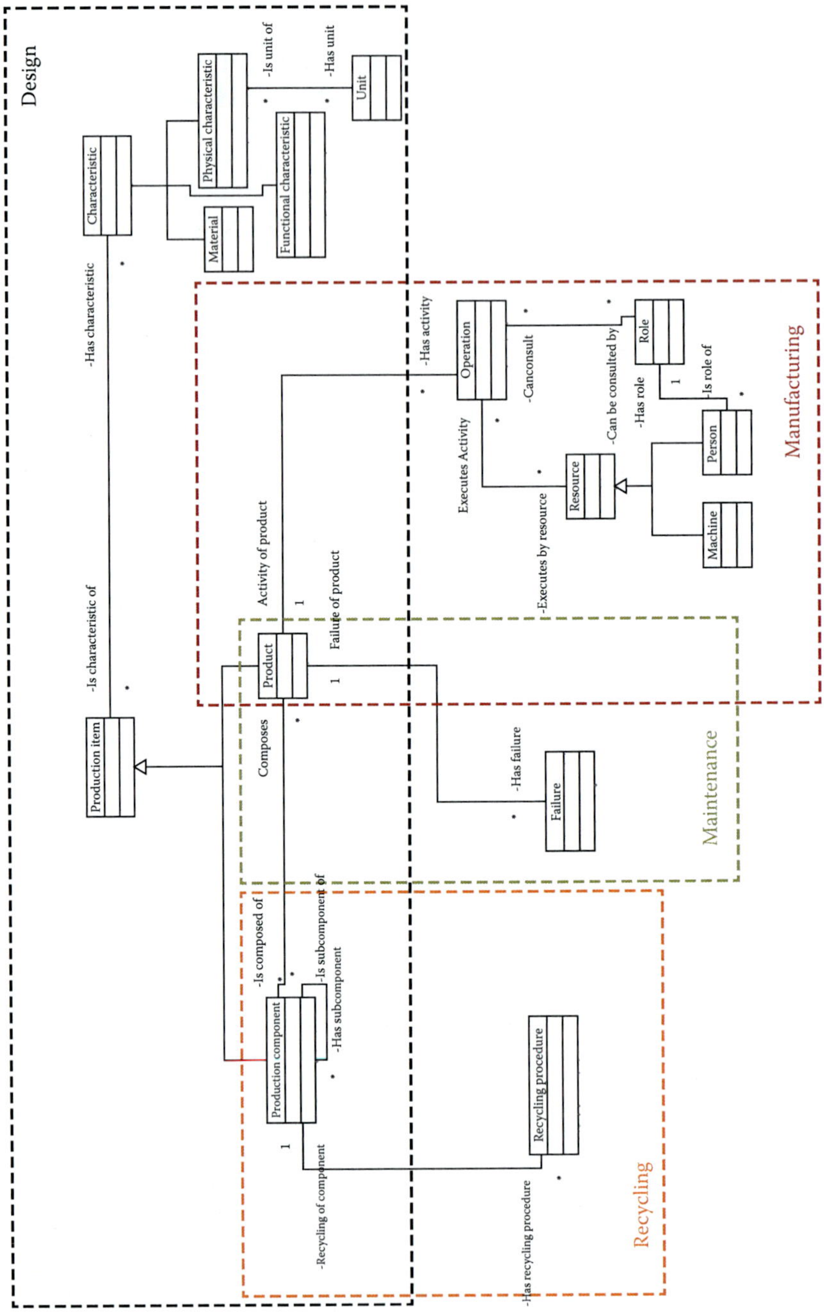

Figure 8.8 UML class diagram of the reference product life cycle ontology.

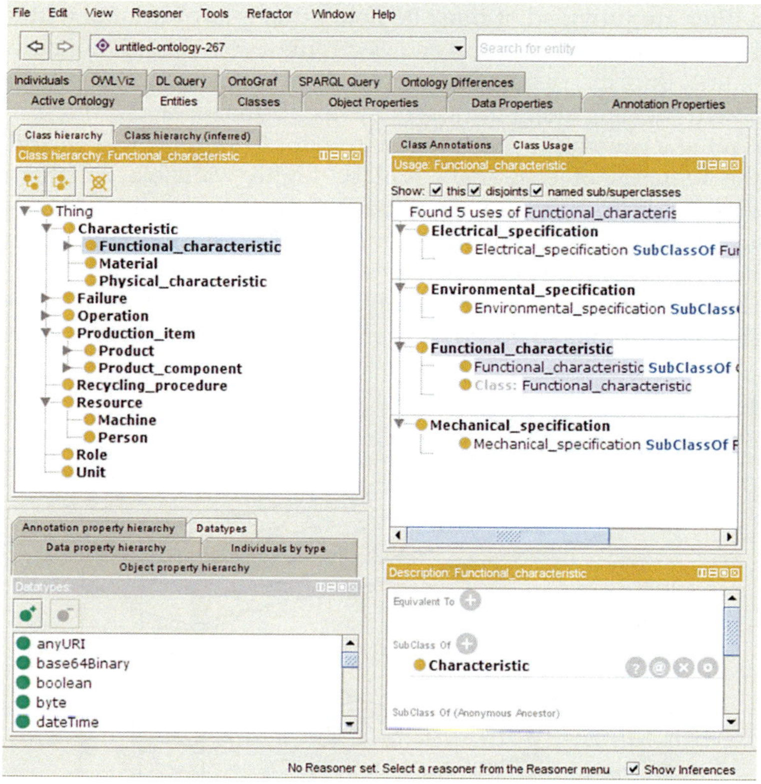

Figure 8.9 Implementation of the reference product life cycle ontology in Protégé.

8.5 Use case

The use case regards a company named RTT, a manufacturer of precision microwave and coaxial cable assemblies, harnesses, and looms in the electronics and telecommunication industry. Their activities involve mechanical assembly, filters for mobile telephony, and the complete assembly of racks, junction boxes, and electronic equipment, especially for aerial or naval means, for civil and military use.

The use case revolves around a telecommunications filter unit, a variant of the RF filter family of units, a technical drawing of which is depicted in Figure 8.10. A telecommunications filter is a product used to send radio signals, such as those used in broadcasting radio, television, and wireless communications. The filter aspect of the unit is to filter out particular frequencies, transmitted or received via the tuning of the product.

RF filter assembly and tuning can be a complex process. While production tends to follow serial lines and the standard assembly steps for any given filter are provided, it must be noted that many variants within the family of filters and mechanical and electronic assemblies are concurrently on the line, thus increasing the complexity of sequencing and scheduling along with tracking along the assembly line (Bruno et al., 2015).

Having a generative product configurator to support RTT in addressing the requests of customers will be highly valuable. Furthermore, knowing in advance the impact of the choices made by clients during the product design phase on the subsequent phases of the product life cycle will help in organizing the company's activities.

The definition of the filter's life cycle is managed closely with the client. The clients exploit their own experience in the design and engineering

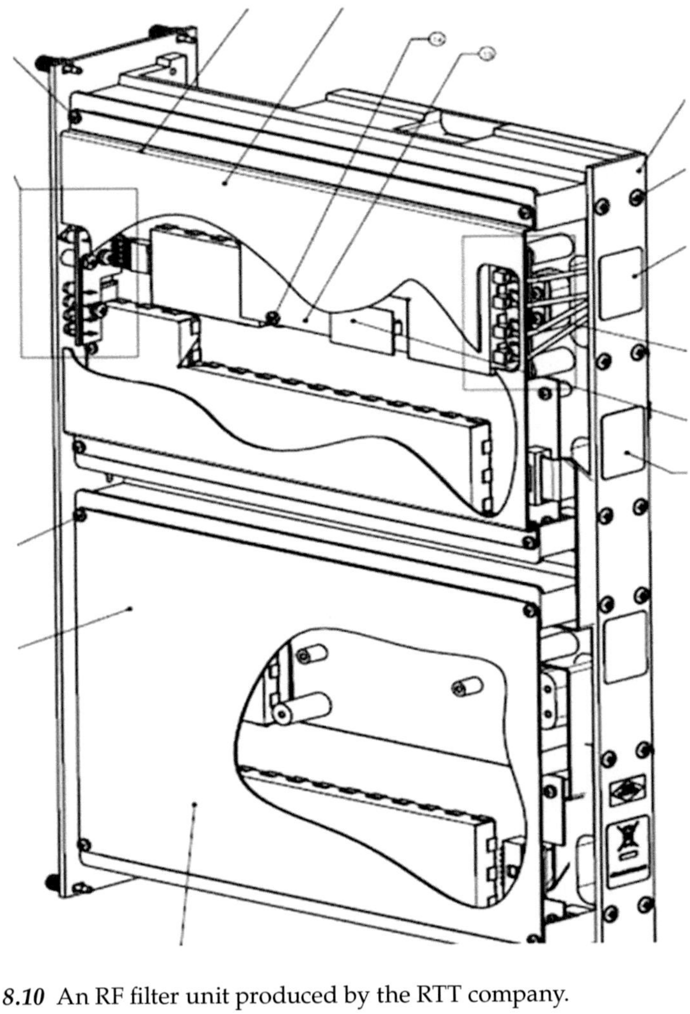

Figure 8.10 An RF filter unit produced by the RTT company.

phases, while relying on the execution capabilities of the company for the filter assembly and tuning.

The design of the filter is done by an Italian enterprise, which then sends the order to assemble a specific filter, identified by its code (e.g., E15R01). During the design phase, the files needed to perform the assembly of the filter (i.e., the BOM, the CAD, and the assembly sequence for the filter and its components), together with the other files needed to tune and test the filters, are generated. By analyzing these files, it is possible to derive the information required to create the ontology classes of the design and manufacturing phases of the life cycle.

After testing the filters, it is possible to report the failures causing filter malfunctions, which are stored in text files. This information is used to derive the ontology concepts for the *maintenance* phase. By analyzing the documents generated through the filter's life cycle, the general concepts are specified into concepts related to the specific filter's life cycle.

In particular, the product component class represents the hierarchy of components that can be included in a filter (e.g., panel, plate, capacitor, inductor, resistor, resonator, and fixing elements), and the functional characteristics represent the characteristics required to produce a filter, divided into electrical specifications (e.g., power supply, frequency range), mechanical specifications (e.g., DC connector, forward and reverse RC connector), and environmental specifications (e.g., operating temperature, relative humidity).

The failure concept is specified into the three main types of failures occurring with a filter: assembly failure, interference, and damage.

The specification of these concepts is graphically represented in Figure 8.11, and the screenshot of the Protégé implementation is shown in Figure 8.12.

By exploiting the structure of information reported in the ontology, on the one hand the clients of RTT will be able to directly design a TCL filter with the desired characteristics, and on the other hand RTT will be able to track the effect of the filter characteristics required by the clients on the production and failures of the filters.

8.6 Conclusions

A large number of industrial sectors benefit from mass customization strategy and are unable to support it with an automatic product configurator. These sectors are constrained by the requirements of a set of product alternatives far too large to be enumerable and determined in advance.

The aim of this chapter is to describe the evolution of product configurator design in the last few decades by highlighting the new trends in exploiting ontologies to define a common vocabulary and integrate data from different sources.

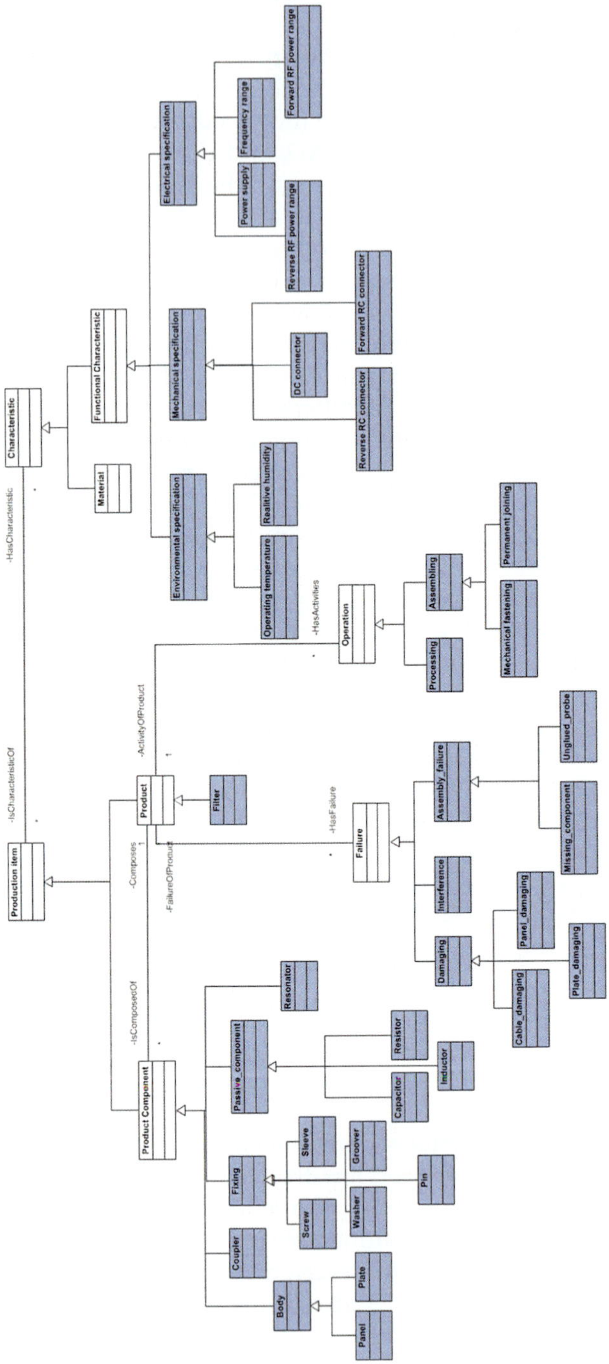

Figure 8.11 UML class diagram containing the specification of the basic concepts for RTT.

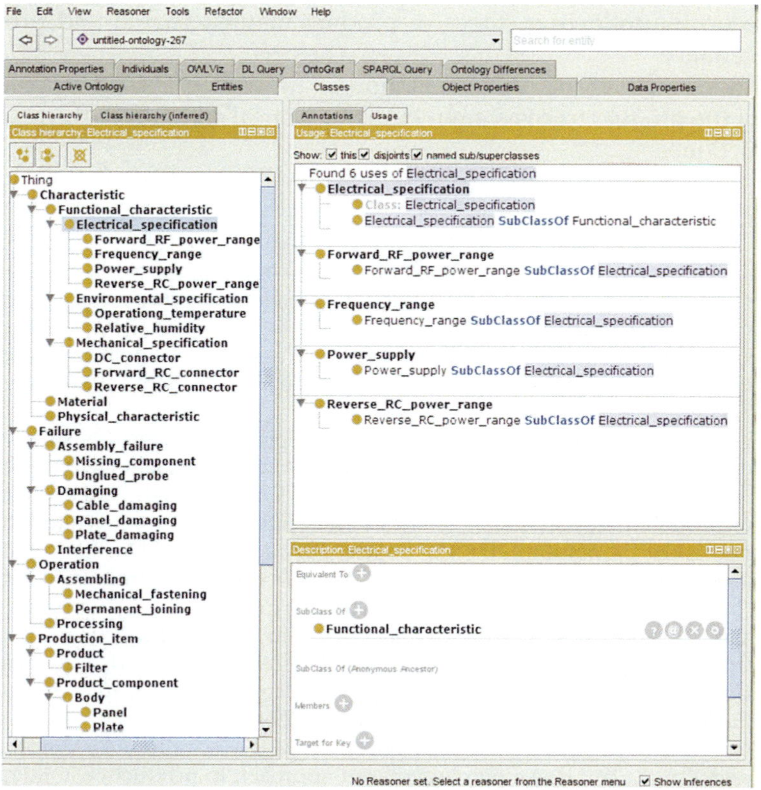

Figure 8.12 Implementation of the RTT ontology in Protégé.

The first evolution of the product configurator was the substitution of lists of product variants with a metamodel of a generic product family usable at the industrial level. The metamodel enables the product configurator to assist the customer during the definition of the product. The metamodel gives the rules to build every possible product tree for every product variant in the family. Since the end of the last century, concepts taken from the object-oriented approach have been applied to design the metatree of the product family.

A further step has been the adoption of ontologies to define not only the part variants that constitute the product but also the different kinds of relationships among the parts.

Presently, product configurators have been integrated into PLM logic because there is a need to add information deriving from the later stages of the product life cycle. Information can thus be exploited in order to evaluate the impact of the choices made during the product design phase on the following stages of product manufacturing and product maintenance.

An example of the implementation of an ontology-based representation of the product tree is shown in the industrial scenario of the RF filter production.

References

Abdul-Ghafour, S. (2012). Integration of product models by ontology development. In *IEEE International Conference on Information Reuse and Integration*, pp. 548–555. IEEE.

Agard, B., and Tollenaere, M. (2002). Conception d'assemblages pour la customisation de masse, Labora-toire GILCO, Grenoble, France. *Mécanique & Industries*, 3, 113–119.

Anderson, C., and Zeithaml, C. (1984). Stage of the product life cycle, business strategy, and business performance. *Academy of Management Journal*, 27(1), 5–24.

Antonelli, D., Bruno, G., Schwichtenberg, A., and Villa, A. (2014). Full exploitation of product lifecycle management by integrating static and dynamic viewpoints. In Emmanouilidis, C., Taisch, M., and Kiritsis, D. (Eds.) *Advances in Production Management Systems, Competitive Manufacturing for Innovative Products and Services*, pp. 176–183. Berlin: Springer.

Antonelli, D., and Villa, A. (2006). Object-oriented design of a generative product configurator. In Monostori, L., and Ilie-Zudor, A. E. (Eds.) *International Conference on the Modern Information Technology in the Innovation Processes of the Industrial Enterprises (MITIP2006)*, pp. 529–534.

Ardito, C., Barricelli, B. R., Buono, P., Costabile, M. F., Piccinino, A., Valtolina, S., and Zhu, L. (2011). An ontology-based approach to product customization. *Lecture Notes in Computer Science*, 6654, 92–106,

Barbarisi, O., Del Vecchio, C., Glielmo, L., and Vasca, F. (2004). Why adopting UML to model hybrid manufacturing systems? *International Conference on Systems, Man and Cybernetics*. IEEE.

Bruno, G., Antonelli, D., Korf, R., Lentes, J., and Zimmermann, N. (2014). Exploitation of a semantic platform to store and reuse PLM knowledge. In Grabot, B., Vallespir, B., Gomes, S., Bouras, A., and Kiritsis, D. (Eds.) *Advances in Production Management Systems*, IFIP Advances in Information and Communication Technology, Vol. 438, pp. 59–66. Berlin: Springer.

Bruno, G., Antonelli, D., and Villa, A. (2015). A reference ontology to support product lifecycle management, *Procedia CIRP*, 33, 41–46.

Catalano, C. E., Camossi, E., Ferrandes, R., Cheutet, V., and Sevilmis, N. (2009). A product design ontology for enhancing shape processing in design workflows. *Journal of Intelligent Manufacturing*, 20, 553–567.

Demartini, C., Rivoira, S., and Valenzano, A. (1998). Product data exchange using STEP. In Jacucci, G., Olling, G. J., Preiss, K., and Wozny, M. J. (Eds.) *Proceedings of the Tenth International IFIP* WG 5.2/5.3, pp. 257–269.

Eckstein, H., and Josefiak, F. (2008). Increased efficiency in customer involvement in configuration processes – The SWOP approach. In *2008 IEEE International Technology Management Conference* (ICE), pp. 1–8.

Eden, A. (2002). A theory of object-oriented design. *Information Systems Frontier*, 4(4), 379–391.

Farrell, R. S., and Simpson, T. W. (2003). Product platform design to improve commonality in custom products. *Journal of Intelligent Manufacturing*, 14(6), 541–556.

Fiorentini, X., Gambino, I., Liang, V. C., Foufou, S., Rachuri, S., Bock, C., and Mani, M. (2007). Towards an ontology for open assembly model. In *International Conference of Product Lifecycle Management*, pp. 445–456.

Forza, C., and Salvador, F. (2002). Managing for variety in the order acquisition and fulfilment process: The contribution of product configuration systems. *International Journal of Production Economics*, 76, 87–98.

Fowler, M., and Scott, K. (2000). *UML Distilled*. Reading, MA: Addison-Wesley.

Grady, P., and Liang, W.-Y. (1998). An object-oriented approach to design with modules. *Computer Integrated Manufacturing Systems*, 11(4), 267–283.

He, D., Kusiak, A., and Tseng, T. (1998). Delayed product differentiation: A design and manufacturing perspective. *Computer-Aided Design*, 30(2), 105–113.

Hedin, G., Ohlsson, L., and McKenna, J. (1998). Product configuration using object oriented grammars. In Magnusson, B. (Ed.) *System Configuration Management*, pp. 107–126. Berlin: Springer.

Hsuan, J. (1999). Impacts of supplier–buyer relationships on modularization in new product development. *European Journal of Purchasing and Supply Management*, 5, 197–209.

Huang, C. C., and Kusiak, A. (1998). Modularity in design of products and systems. *IEEE Transactions on Systems, Man and Cybernetics, Part A: Systems and Humans*, 28, 66–77.

Imran, M., and Young, B. (2013). The application of common logic based formal ontologies to assembly knowledge sharing. *Journal of Intelligent Manufacturing*, 26(1), 139–158.

Jiao, J., Ma, Q., and Tseng, M. M. (2003). Towards high value-added products and services: Mass customization and beyond. *Technovation*, 23, 809–821.

Jiao, J., and Tseng, M. M. (1999). A methodology of developing product family architecture for mass customization. *Journal of Intelligent Manufacturing*, 10, 3–20.

Jurisica, I., Mylopoulos, J., and Yu, E. (2004). Ontologies for knowledge management: An information systems perspective. *Knowledge and Information Systems*, 6, 380–401.

Kwak, J. A., and Yong, H. S. (2008). An approach to ontology-based semantic integration for PLM object. In *IEEE International Work on Semantic Computing and Applications*, pp. 19–26. IEEE.

Lu, T., Guan, F., Gu, N., and Wang, F. (2008). Semantic classification and query of engineering drawings in the ship building industry. *International Journal Production Research*, 46(9), 2471–2483.

Lutters, D., Mentink, R. J., Van Houten, F., and Kals, H. (2001). Workflow management based on information management. *CIRP Annals*, 50(1), 309–312.

Markham, T., Frohlich, A., and Westbrook, R. (2001). Arcs of integration: An international study of supply chain strategies. *Journal of Operations Management*, 19, 185–200.

Martin, G., Lavagno, L., and Louis-Guerin, J. (2001). Embedded UML: A merger of real-time UML and co-design. In *Proceedings of the Ninth International Symposium on Hardware/Software Codesign*, pp. 23–28. IEEE.

Matsokis, A., and Kiritsis, D. (2010). An ontology-based approach for product lifecycle management. *Computers in Industry*, 61, 787–797.

McKenzie-Veal, D., Hartman, N. W., and Springer, J. (2010). Implementing ontology-based information sharing in product lifecycle management. Academia.edu. Elsevier.

Meyer, M. H., and Dhaval, D. (2002). Managing platform architectures and manufacturing processes for nonassembled products. *Journal of Product Innovation Management*, 19(4), 277–293.

Muffatto, M. (1999). Introducing a platform strategy in product development. *International Journal of Production Economics*, 60, 145–153.

Olsen, K. A., Sætre, P., and Thorstenson, A. (1997). A procedure-oriented generic bill of materials. *Computers and Industrial Engineering*, 32(1), 29–45.

Panetto, H., Dassisti, M., and Tursi, A. (2012). ONTO-PDM: Product-driven ONTOlogy for product data management interoperability within manufacturing process environment. *Advanced Engineering Informatics Archive*, 26(2), 334–348.

Sabin, D., and Rainer, W. (1998). Product configuration frameworks: A survey. *IEEE Intelligent Systems*, 4, 42–49.

Sim, S. K., and Duffy, A. H. B. (2003). Towards an ontology of generic engineering design activities. *Research in Engineering Design*, 14, 200–223.

Slimani, K., Silva, C., Ferreira, D., Médini, L., and Ghodous, P. (2006). Conflict mitigation in collaborative design. *International Journal of Production Research*, 44(9), 1681–1702.

Stark, J. (2005). *Product Lifecycle Management: 21st Century Paradigm for Product Realization*. London: Springer.

Tiihonen, J., Soininen, T., Männistö, T., and Sulonen, R. (1996). State of the practice in product configuration: A survey of 10 cases in the Finnish industry. *Knowledge Intensive CAD*, 95–114.

von der Beeck, M., Braun, P., Rappl, M., and Schröder, C. (2003). Automotive UML: A (meta) model-based approach for systems development. In L. Lavagno, G. Martin, and B. Selic, Eds., *UML For Real: Design of Embedded Real-Time Systems*, pp. 271–299. Norwell, MA: Kluwer Academic.

Wang, Q., Peng, W., and Yu, X. (2010). Ontology-based geometry recognition for STEP. In *IEEE International Symposium on Industrial Electronics*, pp. 1686–1691.

chapter nine

Shoe configurators

A comparative analysis of capabilities and benefits

Enrico Sandrin, Cipriano Forza, Zoran Anisic, Nikola Suzic, Chiara Grosso, Thomas Aichner, and Alessio Trentin

Contents

ABSTRACT
Mass customizers (MCs) increasingly sell their products on the web through web-based sales configurators (WBSCs). This selling approach has proved beneficial to both MCs and their customers because, on the one hand, it facilitates the customization process and, on the other hand, it provides a real-time preview of the customized product. However, selling through WBSCs is challenging. Different WBSCs have different capabilities and, consequently, customers perceive different levels of benefits from both the configured products and the customization experience. The present work performs an analysis of state-of-the-art WBSCs for shoes and compares them with other fashion WBSCs in order to help companies and researchers to adopt or develop innovative approaches to enhancing WBSCs.

9.1 Introduction

The vast majority of MCs rely on WBSCs to sell their products online (Fogliatto et al., 2012). This selling approach has proved beneficial to both MCs (Heiskala et al., 2007; Forza and Salvador, 2008; Trentin et al., 2011, 2012) and their customers (Grosso et al., 2014; Trentin et al., 2014; Franke et al., 2010). However, selling through WBSCs is challenging, not only because it is a new way of selling for many companies, but also because WBSCs are witnessing a number of continuous technical innovations. This challenge can be observed across different sectors, but it is particularly evident in the fashion industry and, more precisely, in the footwear industry. The trend to mass customize shoes is in line with the exponential increase in product variety of this product category. For example, in the United States there has been an increase from five sport shoe models in 1970 to 285 models in 1998 to 3371 models in 2012 (Aichner and Coletti, 2013).

The present work performs an analysis of the state-of-the-art WBSCs for shoes in order to help companies and researchers to adopt or develop innovative solutions to support online product customization. More specifically, it analyzes WBSCs both for shoes and for other fashion-related products. The analysis identifies improvement opportunities and best practices for shoe configurators to augment the value perceived by customers from shoe personalization.

Evaluating product configurators is a complex task. In order to ensure that the present research is rigorous, the assessment is based on the WBSC capabilities proposed by Trentin et al. (2013). The capabilities under consideration (user-friendly product space description, focused navigation, flexible navigation, benefit–cost communication, and easy comparison) have been proposed to reduce the difficulties a customer faces when he or she customizes a product (Trentin et al., 2013). However, sales configurators can act not only as tools to reduce customer difficulties but also as a means to increase the benefits that customers derive from customization. In that respect, the benefits identified by Merle et al. (2010), that is, utilitarian, uniqueness, self-expressiveness, hedonic, and creative achievement benefits, are taken into account.

Based on the measures proposed and tested by Merle et al. (2010), Trentin et al. (2013, 2014), and Grosso et al. (2014), both shoe configurators and configurators of other fashion products are evaluated. The 68 web-based configurators were evaluated by a total of 98 users. Each configurator was evaluated, on average, by five different users. The 333 different configuration experiences have been analyzed to identify the strengths and weaknesses of WBSCs for shoes. Subsequently, the shoe configurators with the highest average evaluations have been further analyzed to identify the solutions that those configurators use to achieve each capability. Finally, a number of nonshoe configurators with high capabilities have been analyzed to identify solutions that shoe configurators could adopt to improve those capabilities on which they are lower.

9.2 Product configurator capabilities

Drawing on prior research concerning sales configurators and the customer decision process, Trentin et al. (2013) defined five capabilities that sales configurators should deploy in order to alleviate the risk that offering more product variety and customization in an attempt to increase sales paradoxically results in a loss of sales: (1) a user-friendly product space description, (2) focused navigation, (3) flexible navigation, (4) benefit–cost communication, and (5) easy comparison capabilities.

9.2.1 *User-friendly product space description capability*

User-friendly product space description capability is the ability of a sales configurator to adapt the description of a company's product space to the individual characteristics of a potential customer as well as to the situational characteristics of his or her use of the sales configurator (Trentin et al., 2013, 2014). The essence of this capability is captured, for example, by the following statements, which were used, with few modifications, as measures in Trentin et al. (2014).

- The system gives an adequate presentation of the choice options for when the user is in a hurry, as well as when the user has enough time to go into the details.
- The product features are adequately presented for the user who just wants to find out about them, as well as for the user who wants to go into specific details.
- The choice options are adequately presented for both the expert and inexpert user of the product.

An example of a WBSC deploying this capability is on Volkswagen's website (www.volkswagen.co.uk/configurator). This configurator allows potential customers to customize Polo utility cars and, for each available choice, enables users to opt for either a brief description or a detailed one with more technical information, which is available by selecting the "More Info" button. In addition, choices affecting the aesthetics of the car are described using both text and product images, which change automatically as the potential customer selects different options.

9.2.2 *Focused navigation capability*

Focused navigation capability is the ability of a sales configurator to quickly focus a potential customer's search on those solutions of a company's product space that are most relevant to the customer, such as those that are most likely to satisfy his or her idiosyncratic needs (Trentin et al., 2013, 2014). The essence of this capability is captured, for example, by the following statements, which were used as measures in Trentin et al. (2014, p. 702).

- The system made me immediately understand which way to go to find what I needed.

- The system enabled me to quickly eliminate from further consideration everything that was not interesting to me at all.
- The system immediately led me to what was more interesting to me.
- This system quickly leads the user to those solutions that best meet his or her requirements.

An example of a WBSC with this capability is Lenovo's laptop customization website (www.lenovo.com). This configurator has a search engine called Laptop Finder, which enables potential customers to narrow their search to laptops with a specific color, processor, and/or any other customizable feature of the product.

9.2.3 *Flexible navigation capability*

Flexible navigation capability is the ability of a sales configurator to let its users easily and quickly modify a product configuration that they have previously created or one that they are currently creating (Trentin et al., 2013, 2014). The essence of this capability is captured, for example, by the following statements, which were used, with few modifications, as measures in Trentin et al. (2014).

- The system enables the user to change some of the choices the user has previously made during the configuration process without having to start it over again.
- With this system, it takes very little effort to modify the choices the user has previously made during the configuration process.
- Once the user has completed the configuration process, this system enables the user to quickly change any choice made during that process.

An example of a WBSC deploying this capability is on Converse's website (www.converse.com/us). This configurator allows potential customers to customize their sports shoes through a multistep configuration process, where each step corresponds to one customizable feature of the product. This process is depicted by a progress bar with a box for each step. Potential customers can modify any previously selected feature simply by clicking on the related box, without losing any other choice that they have made before.

9.2.4 *Benefit–cost communication capability*

Benefit–cost communication capability is the ability of a sales configurator to effectively communicate the consequences of the configuration choices made by a potential customer, both in terms of what he or she would get and in terms of what he or she would give (Trentin et al., 2013, 2014). The essence of this capability is captured, for example, by the following statements, which were used as measures in Trentin et al. (2014, p. 702).

- Thanks to this system, I understood how the various choice options influence the value that this product has for me.
- Thanks to this system, I realized the advantages and drawbacks of each of the options I had to choose from.
- This system made me exactly understand what value the product I was configuring had for me.

An example of a WBSC with this capability is Dell's laptop customization website (www.dell.com). At each step of the configuration process, this site gives potential customers the possibility to click on the "Help Me Choose" button, which opens up a page with a list of recommendations outlining the advantages of every single option. Furthermore, the site communicates the price variations that selecting each of the available options would cause with respect to the price of the current configuration.

9.2.5 *Easy comparison capability*

Easy comparison capability is the ability of a sales configurator to support its users in comparing product configurations they have previously created (Trentin et al., 2013, 2014). The essence of this capability is captured, for example, by the following statements, which were used, with few modifications, as measures in Trentin et al. (2014).

- The system enables easy comparison of product configurations previously created by the user.
- The system allows the user to easily understand what previously created configurations have in common.
- The system enables side-by-side comparison of the details of previously saved configurations.
- The system allows the user to easily understand the differences between previously created configurations.

An example of a WBSC that deploys this capability is on Nike's website (www.nike.com). This configurator enables potential customers to save configured shoes in their "Wish List" area and to subsequently access them at any time. Once logged in, the site user finds all of his or her previously saved shoes portrayed on the same web page, so they can be easily compared.

9.3 Consumer-perceived benefits

As recently acknowledged in the literature, mass customization involves not only improving compatibility between product customization and a firm's operational performance (Pine, 1993; Tu et al., 2001; McCarthy, 2004), but also augmenting the customer's perceived benefits with regard to both the customized products and the customization experience (Schreier, 2006; Merle et al., 2010; Fogliatto et al., 2012). Prior research on the consumer-perceived value of a mass customized product (e.g., Merle et al., 2010) explains that in addition to the well-researched utilitarian benefit, there are two benefits that a consumer could derive from the possession of a mass customized product—namely, uniqueness and self-expressiveness. Mass customization research has also identified two other benefits—namely, a creative achievement benefit and a hedonic benefit, which a consumer derives not from the possession of a mass customized product, but from the experience of self-customizing such a product using a sales configurator (e.g., Merle et al., 2010).

9.3.1 Utilitarian benefit

The utilitarian benefit is a benefit derived from the closeness of fit between objective product characteristics (i.e., physical, aesthetic, functional characteristics) and individual preferences about the product's functional/instrumental characteristics (Merle et al., 2010). Therefore, the utilitarian benefit deriving from a mass customized product fulfills the consumer's functional or aesthetic needs related to a self-designed product (Addis and Holbrook, 2001). To measure this benefit, Grosso et al. (2014, p. 86), building on Merle et al. (2010), used the following three statements:

- This product is exactly what I had hoped for.
- I was able to create a product that was the most adapted to what I was looking for.
- I was able to create the product I really wanted to have.

9.3.2 Uniqueness benefit

The uniqueness benefit of possessing a mass customized product is defined as the benefit that a consumer derives from the opportunity to assert his or her personal uniqueness by possessing a customized product (Merle et al., 2010). The uniqueness benefit is related to the symbolic meanings a person attributes to objects as a result of social construction. Brewer's optimal distinctiveness theory posits that people have opposing motives: to fit in and to stand out from social groups (Brewer, 1991). Whereas threats to one's inclusionary status produce increased attempts to fit in and conform, threats to one's individuality produce attempts to demonstrate how different one is from the rest of the group. Consequently, the uniqueness benefit deriving from a mass customized product meets the individual need to assert his or her own personality by differentiating himself or herself from others. To measure this benefit, Grosso et al. (2014, p. 86), building on Merle et al. (2010), used the following three statements:

- With this product, I will not look like everybody else.
- With this program, I was able to design a product that others will not have.
- With this product, I have a small element of differentiation compared with others.

9.3.3 *Self-expressiveness benefit*

The self-expressiveness benefit is the benefit that originates from the opportunity to possess a product that is a reflection of the consumer's self (Merle et al., 2010). This is in accordance with the self-consistency motive underlying self-concept, where the term *self-consistency* denotes the tendency for an individual to behave consistently with his or her view of himself or herself. Like uniqueness, the self-expressiveness benefit is related to the symbolic meanings a person attributes to objects as a result of social construction. Possessions are often an extension of the self. As Belk (1988) stated, "People seek, express, confirm, and ascertain a sense of being through what they have" (p. 146). This statement implicitly relates identity with consumption. Consumers deliberately acquire things and engage in consumption practices to achieve a preconceived notion of their selves. Thus, a mass customized product will fulfill individual needs for self-consistency by offering possession of a product that is a reflection of the consumer's self. To measure this benefit, Grosso et al. (2014, p. 87), building on Merle et al. (2010), used the following three statements:

- I was able to create a product that is just like me.
- This product reflects exactly who I am.
- This product is in my own image.

9.3.4 *Creative achievement benefit*

The creative achievement benefit derives from the capacity of the mass customization experience to arouse pride of authorship. In general, pride is a positive emotion of self-reward that follows the assessment of one's competence in a situation that is, in some measure, challenging, such as an exam or climbing a mountain. Pride of authorship, in particular, is the feeling of pride that an individual experiences whenever he or she creates, or at least has a sense of being the creator of, an artifact that constitutes positive feedback on his or her own competencies. To measure this benefit, Trentin et al. (2014, p. 703), building on Merle et al. (2010), used the following three statements:

- I felt really creative while configuring this product.
- The company gave me a lot of freedom while creating this product.
- By personalizing this product, I had the impression of creating something.

9.3.5 *Hedonic benefit*

Unlike the creative achievement benefit, the hedonic benefit stems only from the characteristics of the mass customization experience and, therefore, may be enjoyed even though a potential customer does not complete the configuration task. Hedonic benefit derives from the capacity of the mass customization experience to be intrinsically rewarding. There may be various reasons why a generic activity can be an end in itself, thus implying the actor's positive affect (enjoyment, contentment, satisfaction, etc.). To measure this benefit, Trentin et al. (2014, p. 703), building on Merle et al. (2010), used the following three statements:

- I found it fun to customize this product.
- Customizing this product was a real pleasure.
- Customizing this product was like a game.

9.4 *Method*

To assess the state-of-the-art web-based shoe configurators, a set of 333 customization experiences from 98 users on a sample of 68 WBSCs was used. These experiences allowed the assessment of both the WBSC capabilities and the benefits presented in Sections 9.2 and 9.3, respectively. This analytical approach has been successfully adopted by Merle et al. (2010), Trentin et al. (2013, 2014), and Grosso et al. (2014).

The sample of 68 WBSCs used in the present study includes both shoe configurators and other fashion-related product configurators. Each configurator has been evaluated, on average, through five different experiences, each one performed by a different user. In each experience, one of the 98 users browsed an assigned sales configurator website and configured a product according to his or her own preferences. At the end of each experience, the user filled out a questionnaire developed to measure the capabilities and benefits presented in the previous sections. In this questionnaire, the measures proposed and tested by Merle et al. (2010), Trentin et al. (2013, 2014), and Grosso et al. (2014) were used. These measures

Table 9.1 Sample of WBSCs

Product category	No. of web configurators	No. of observations
Men's and women's shoes	5	32
Sneakers, sports shoes	7	46
Total shoes	12	78
Suits and elegant men's clothing, women's clothing, wedding dresses	4	24
Wedding rings, watches, women's jewelry	14	49
Bags, messenger bags, women's handbags, backpacks	12	54
Shirts	9	53
T-shirts, jeans, shorts, and so on	19	75
Total other fashion products	57*	255
Total	68*	333

* These numbers are lower than the sums per category because one configurator sells products belonging to more than one product category.

were presented, with few modifications, in Sections 9.2 and 9.3. Table 9.1 shows the product categories involved and, for each category, the number of different WBSCs and the number of mass customization experiences (observations).

9.5 *Benchmarking: Capabilities*

For each capability, a calculation was done of the average score within the 12 shoe configurators (product categories of "Men's and women's shoes" and "Sneakers, sports shoes") and the average score within the 57 other fashion product configurators (all the other product categories except shoes). The results are reported in the radar chart in Figure 9.1. This chart allows comparison of the various capabilities of shoe configurators to identify the more and less advanced capabilities. It also allows comparison of shoe and nonshoe fashion product configurators to assess whether shoe configurators are more or less advanced in comparison with other fashion product configurators. The scale used in Figure 9.1 is the same as the response scale used in the questionnaire. When responding to the questionnaire, respondents specified their level of agreement or disagreement on a 1–7 symmetric agree–disagree scale, where 1 means "Completely agree," 4 means "Neither agree nor disagree," and 7 means "Completely disagree," for the set of statements used to measure each capability. The final score for each capability is the simple mean of the capability statement scores.

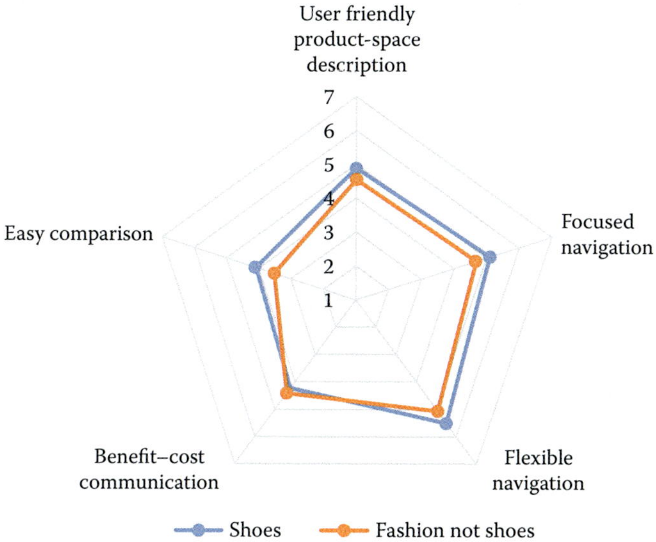

Figure 9.1 Benchmarking of WBSC capabilities.

In absolute terms, shoe configurators feature the following levels of the five capabilities:

- High on flexible navigation and focused navigation
- Intermediate–high on user-friendly product space description
- Intermediate–low on benefit–cost communication
- Low on easy comparison

Compared with nonshoe fashion product websites, the scores for WBSCs for shoes were as follows:

- Higher for flexible navigation, focused navigation, user-friendly product space descriptions, and easy comparison.
- Slightly lower for benefit–cost communication.

9.6 Benchmarking: Benefits

For each benefit, a calculation was done of the average score within the 12 shoe configurators (product categories of "Men's and women's shoes" and "Sneakers, sports shoes") and the average score within the 57 other fashion product configurators (all the other product categories except shoes). The results are reported in the radar chart in Figure 9.2. This chart compares shoe WBSCs and other fashion WBSCs on product-related and customization experience–related benefits. Prior research

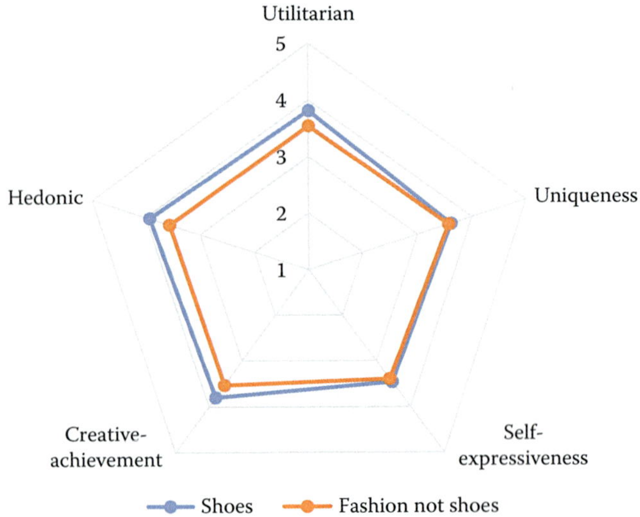

Figure 9.2 Benchmarking of WBSCs on product-related and customization experience–related benefits.

provides empirical evidence that WBSC capabilities influence at least some of these benefits (Grosso et al., 2014; Trentin et al., 2014). Therefore, even though other factors, such as the characteristics of the company's solution space, are likely to play a role in determining such benefits, it is meaningful to compare shoe WBSCs and other fashion WBSCs on these benefits as well. The scale used in Figure 9.2 is the same as the response scale used in the questionnaire. When responding to the questionnaire, respondents specified their level of agreement or disagreement on a 1–5 symmetric agree–disagree scale, where 1 means "Completely agree," 3 means "Neither agree nor disagree," and 5 means "Completely disagree," for the set of statements used to measure each benefit. The final score for each benefit is the simple mean of the scores for the benefits statements on the questionnaire.

In absolute terms, the evaluation of shoe WBSCs on product-related and customization experience–related benefits provides the following levels of the five benefits:

- High levels on hedonic, creative achievement, and utilitarian benefits
- Intermediate–high levels on self-expressiveness and uniqueness benefits

Compared with nonshoe fashion product websites, web-based shoe configurators scored as follows:

- A little higher on hedonic, creative achievement, and utilitarian benefits
- Similar on self-expressiveness and uniqueness benefits

9.7 *Best practices from shoe configurators*

This section presents, for each capability, some of the solutions that the shoe configurators with the highest average evaluations adopt to achieve that capability.

9.7.1 *User-friendly product space description capability*

From the analyses conducted, some of the solutions that are associated with higher user-friendly product space description capability are those implemented by Shoes of Prey and Adidas.

9.7.1.1 Shoes of Prey

The Shoes of Prey website (www.shoesofprey.com) provides a large number of color and material options that allow a user to personalize many details of the shoe. For each element, an explanatory image is provided (e.g., the material, the style, or the type of heel; Figure 9.3). In addition, by holding the mouse cursor over a certain option, the user is provided with a drop-down window with a textual description of the element that is represented. The elements are fairly intuitive, so an image is more than enough to understand what is intended. However, a description of the material is still useful because small product pictures cannot effectively provide this kind of information, leaving the user to guess. The site also provides a detailed explanation that guides the user in navigating through the configurator and, in particular, in choosing shoe sizes. In addition, there is a step-by-step tutorial that guides the user the first time he or she starts the configuration of a product, explaining the main features of the process.

9.7.1.2 Adidas

The Adidas website (www.adidas.com) guides the customer step by step in the customization of a shoe, which can go very deep. In particular, once the user chooses a basic model from which to start the configuration (which is divided into "Men's shoes," "Women," or "Kids"), the website describes the shoe performance in detail, explains in what contexts the shoe ensures better performance, and so on.

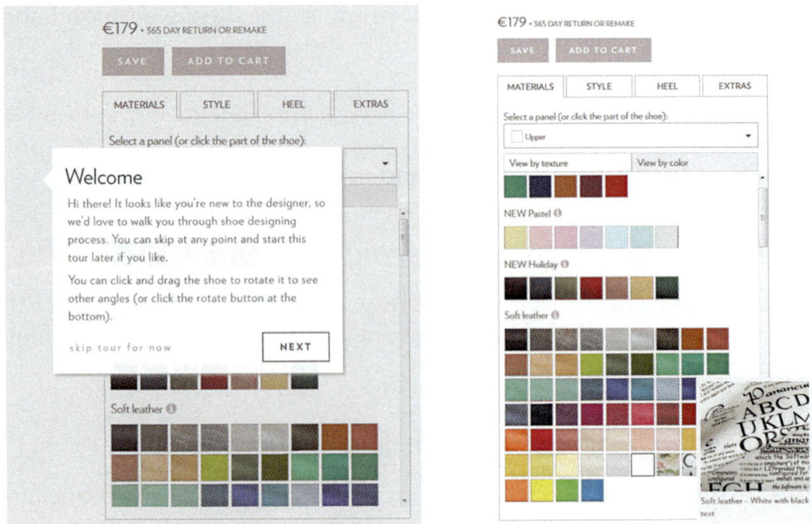

Figure 9.3 The user-friendly product space description capability in the Shoes of Prey WBSC.

A detailed description of alternatives from which to choose is not provided because the choices are very intuitive, relating mostly to the color of the customizable part, for which a representation of the available color variations is sufficient.

Moreover, this configurator even provides a detailed table with shoe sizes in the specific units of measure for different countries to facilitate the user's decision regarding what size to buy.

9.7.2 *Focused navigation capability*

Some of the solutions that are associated with higher focused navigation capability are those implemented by Shoes of Prey and Reebok.

9.7.2.1 *Shoes of Prey*

The Shoes of Prey women's shoes configurator shows, from the beginning of its use, the various ways that users can navigate. Various ways to navigate are specified clearly at the top of the page and, from there, the user can select what he or she wants to do or see. In addition, the configurator menus offer the ability to immediately choose a basic shoe model to start the configuration process, in case the customer is already clear about his or her needs. Many alternatives are offered, including ankle boots, high-heeled shoes, medium-heeled shoes, low-heeled shoes, ballet-style shoes, sandals, open-toed shoes, closed-toed shoes, and so on.

9.7.2.2 Reebok

The Reebok sneakers configurator (www.reebok.com), as well as the Adidas one, typically interfaces with a diverse clientele and supports users in making their choices with mechanisms that allow simplification of the search space of the offered product. For example, the search is simplified through the activation of several customizable filters, making it possible to remove shoe models that do not reflect the search criteria. Then, the customer is directed toward models that might better reflect his or her needs.

An example of these tools is shown in Figure 9.4. The configurator allows the user to narrow the initial set of available shoes to configure by selecting a set of options in the menu on the left and sorting the available shoes by various criteria, such as "Top sellers" or "Price." By means of the menu on the left, the user can select shoe characteristics through various filter options such as "Gender" (e.g., if the shoe is for men, for women, or for kids) and the type and use for which the user is configuring the shoe (e.g., if the shoe is classic, CrossFit, or for running). Subsequently, it is possible to click on the preferred model from the set of shoe models that meet the selected criteria, as the starting point for the configuration process.

Moreover, the configurator directs the customization steps toward the final configuration, presenting one attribute of the product at a time to the customer.

9.7.3 Flexible navigation capability

Some of the solutions that are associated with higher flexible navigation capability are those implemented by Shoes of Prey and Adidas.

9.7.3.1 Shoes of Prey

Shoes of Prey is a very flexible website, since it is easy to change the choices that the user makes during the configuration of the product, and it is equally easy to change a configuration made earlier. Each choice is independent of the others, and it is up to the user to decide how to proceed. For example, the user can decide the order in which he or she defines individual attributes and can change them if they are not in tune with each other or if they are not to the user's liking. The site permits the maximum freedom of choice in configuration, although some materials cannot be used for certain parts of the shoe. The user can also save his or her product configurations to resume them later, without prior registration on the website.

9.7.3.2 Adidas

The Adidas configurator, like the majority of shoe configurators such as Reebok, Nike, and Converse, allows the user to easily change the current

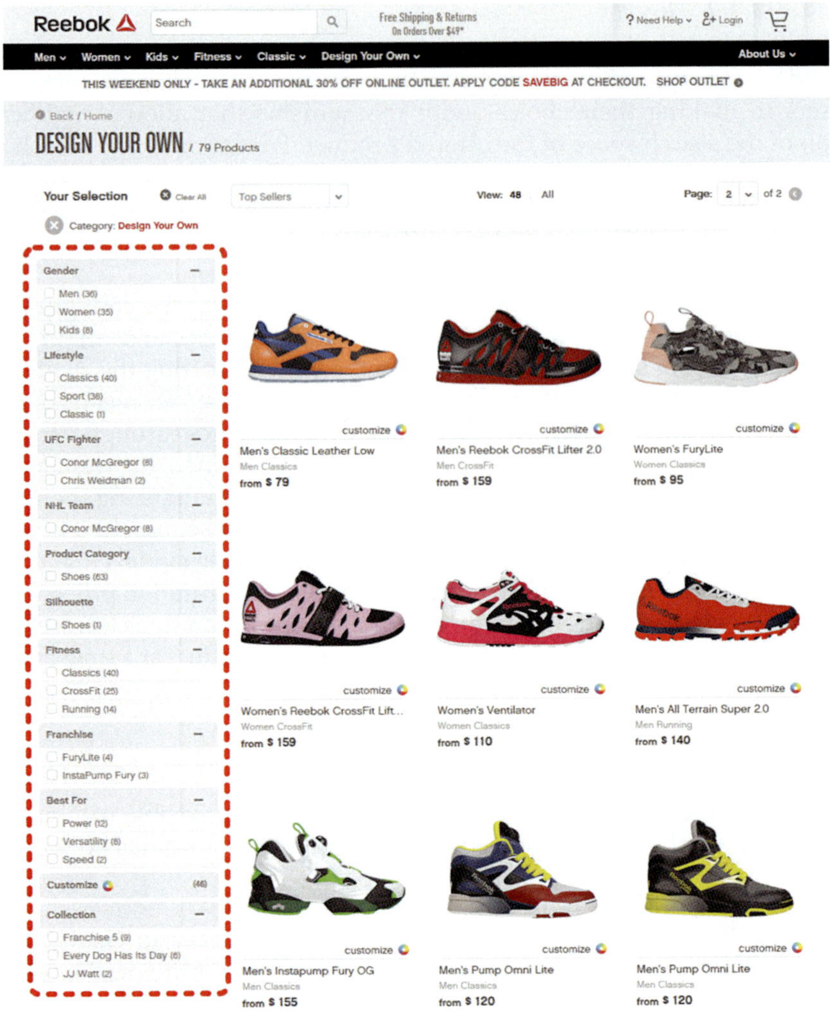

Figure 9.4 The focused navigation capability in the Reebok WBSC.

configuration, returning to previous options through a navigation menu or by selecting a feature directly on the displayed shoe (both highlighted in Figure 9.5), thus changing the previous choices by directly selecting another option. There are no constraints between options belonging to different characteristics. With the Adidas configurator, every potential consumer is able to go back and fairly easily change his or her customization choices, without triggering changes in the parameters that they do not want to change.

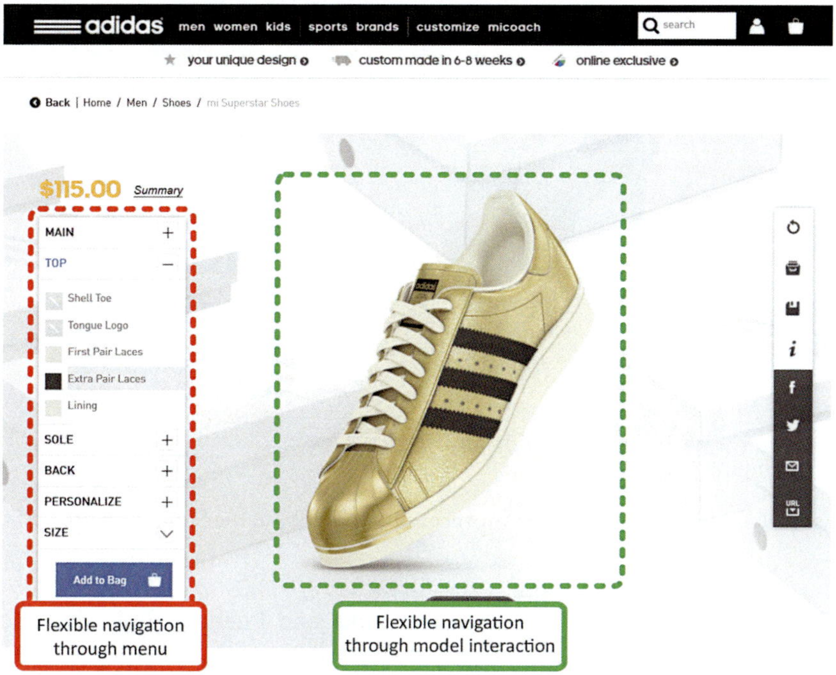

Figure 9.5 The flexible navigation capability in the Adidas WBSC.

9.7.4 Benefit–cost communication capability

Some of the solutions that are associated with higher benefit–cost communication capability are those implemented by Shoes of Prey and New Balance.

9.7.4.1 Shoes of Prey

The Shoes of Prey WBSC clearly and unequivocally presents the features related to each choice and allows the verification of all the costs for each choice in addition to the base price. During the configuration process, the total price is always displayed at the top right of the page and it changes as the various options are selected. The user can see the window that contains the price details by moving the cursor over the price, as shown in Figure 9.6. An explanatory image is provided for each option and, if the image is not sufficiently clear, the user can hold the mouse cursor over the option and textual information appears on a drop-down menu. However, the additional costs associated with each alternative are not shown before or during their selection. Instead, the user has to check the price from time to time during the configuration process.

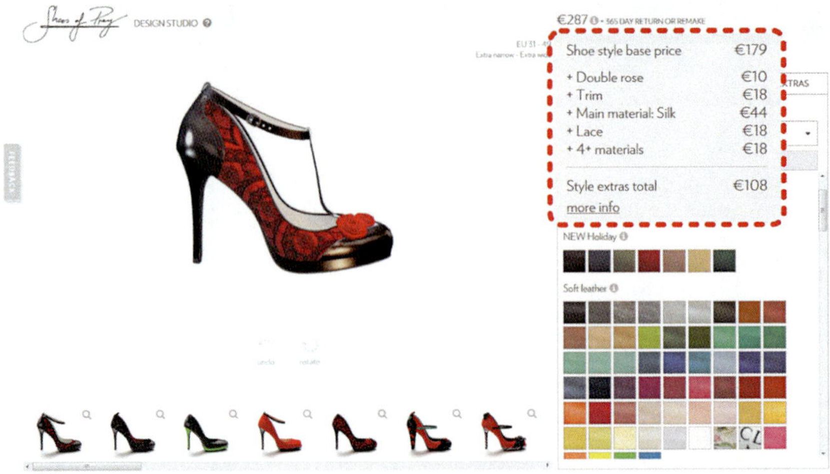

Figure 9.6 The benefit–cost communication capability in the Shoes of Prey WBSC.

Other shoe configurators, such as Nike and Reebok, show the price composition details, specifying the extra costs associated with each selected feature near the virtual image of the customized shoe.

9.7.4.2 *New balance*

New Balance's configurator (www.newbalance.com) adopts a different approach for benefit–cost communication compared with Shoes of Prey and other companies such as Nike, Adidas, and Reebok. The latter four configurators inform the user about the final price and the extra costs associated with each option. Conversely, the price of New Balance's customized shoe is fixed regardless of the type of customization the customer chooses, and it is communicated from the beginning of the configuration process. Therefore, the user knows the price of the selected model of shoes independently from the type of customization. Some users have perceived this characteristic as a good method to effectively communicate benefits and costs, given that during the configuration process there were no consequences of their choices in terms of final price.

Note that configuration is based mostly on choices concerning the color of the different parts of the shoe, which is very intuitive and does not require additional clarification besides an image of the selected color. Nevertheless, there is a section at the bottom of the page that shows the brand history and a description of the particular model selected, with an explanation of the characteristics and materials of which the product is composed.

9.7.5 Easy comparison capability

Some of the solutions that are associated with higher easy comparison capability are those implemented by Shoes of Prey and Nike.

9.7.5.1 Shoes of Prey

Shoes of Prey gives the user an opportunity to compare the configurations that have been created previously by simply placing them in the cart, where they are saved and will remain even if the user leaves the site and returns to it again weeks later. From there, the user can compare products through images of the respective configurations, the prices, and the materials of the product.

Alternatively, the user can create an account to store all of his or her creations and to build a personal collection. Furthermore, the user can add the products that he or he likes the most to a *wish list*, which he or she can access and which gives an overview of all the products he or she has selected. Moreover, the user can compare product characteristics such as the model of shoe, the heel height, the type of texture, the material, and so on.

9.7.5.2 Nike

The Nike site offers the possibility of comparing the results of all the user's configurations on a single page, which, in this case, is constituted by the virtual shopping cart. A comparison of the products is made by placing different types of information about the different customizations of the product side by side. The configurator shows a visual representation of the final results of the configuration process, a textual summary of the configuration choices, and the final price of each product, allowing visual, textual, and price comparisons (Figure 9.7).

9.8 Best practices from nonshoe configurators

This section presents some of the solutions adopted by nonshoe configurators with the highest average evaluations to achieve user-friendly product description capability, benefit–cost communication capability, and easy comparison capability—the three capabilities in which web-based shoe configurators are less advanced.

9.8.1 User-friendly product description capability

9.8.1.1 Audi

A solution that developers of shoe WBSCs could consider to obtain user-friendly product description capability is that adopted on the Audi website (www.audi.com), which uses both a 3-D model of the car at

Figure 9.7 The easy comparison capability in the Nike WBSC.

the center of the screen (which accurately represents the product) and detailed descriptions. For example, the site provides a description of the car's characteristics (rim material, type of brakes, headlights, etc.) and any constraints in the combinations of attributes. Moreover, explanatory photos and videos for the various options that the user can choose from are available, so that even the novice user can gain an idea of the feature and, thus, configure his or her car in a conscious way. The 3-D model of the interior and exterior of the car changes in real time as the customer selects various options. Furthermore, during the configuration of the particular model, a detailed list of the features included in each version is provided.

9.8.1.2 Moja Mix

Another site that developers of shoe WBSCs could consider to obtain this capability is Moja Mix (www.mojamix.com), which offers users the possibility to personalize breakfast cereals based on the customer's own tastes and nutritional needs. The Moja Mix website allows customers to select their ideal cereal mix, composed of a cereal base, dried fruit (cranberries, currants, mango, etc.), nuts and seeds (walnuts, hazelnuts, almonds, etc.), and extras (chocolate chips, chocolate coffee beans, etc.). During the configuration process, the user can choose to simply look at the name of each option and the associated image. If he or she wishes to get more information, however, the user can click on the "More" button highlighted in red and, thus, learn about the product's properties in terms of the nutrient intake associated with the particular choice. These explanations are provided in language comprehensible even to the most inexperienced user, putting the optimal diet choices in understandable terms. Finally, by clicking on the "Nutrition Facts" button, the user who is more experienced in the nutrition field can find a nutritional table explaining in detail the supply of nutrients per serving.

9.8.2 Benefit–cost communication capability

9.8.2.1 Hemdwerk

The Hemdwerk site (www.hemdwerk.com), which sells men's shirts, presents an interesting solution to achieving the benefit–cost communication capability. A 3-D model within the configurator effectively communicates the outcomes of the customization choices. The model provides a 3-D representation of the shirt on the right side of the window, and it changes as the user selects various options. This model helps the user gain a clear idea of how the shirt will look, allowing him or her to see the effects of his or her choices.

The configurator also lets the user know the price of the configured product, which is updated based on the sum of the prices corresponding to the selected components/features. There are additional charges for some options. For example, the addition of a standard pocket button does not involve additional costs, while the addition of a right pocket incurs a small charge. For options that include a specific price, the price is shown clearly beside the option. In this way, when setting up the product, the customer is immediately aware of what features will most increase the final price.

9.8.2.2 NFL shop

In a similar way to Hemdwerk, the NFL Shop (www.nflshop.com), a website where users can purchase customizable sportswear, shows good communication of the prices associated with each of the choices available to the customer. The user is free to modify the product that he or she has selected and a model of the product shown at the top of the site is updated to reflect any changes. Also, the options that can be chosen for the clothing product are at the bottom of the screen, and include various team logos, the range of colors available, and more. All of this is accompanied by the price, which is written on each available option.

An overall summary of the user's choices is progressively updated to the right of the product model, showing all of the cost items that make up the final price in detail, so the user can control the price based on the product options.

9.8.2.3 Moja Mix

Another site that the developers of shoe WBSCs could consider to achieve benefit–cost communication capability is Moja Mix.

The communication of the ingredient's characteristics is shown for each available component of the custom mix, both in terms of the cost of each component and in terms of properties such as its nutritional content. This configurator provides significant support to the potential customer, communicating every available alternative through macrocategories that include both additional costs and information about the beneficial properties of certain foods. For example, the configurator explains the nutritional intake, the details of the products, and what they are recommended for, guiding the user in order to align the choice to his or her needs.

In addition to the mix of cereals that the user is composing, product nutrition information is summarized in a table to assist customers who are less experienced in the field of nutrition. A running list of the chosen ingredients along with the final cost is provided as the customer progresses through the menu. All of this makes the experience pleasant because the user is able to fully understand the product itself and the properties of the product that he or she is creating, and he or she is aware, moment by moment, of the final cost that may incur.

9.8.3 Easy comparison capability

9.8.3.1 Ferrari

An interesting way to implement the easy comparison capability has been proposed by Ferrari, whose website (www.ferrari.com) permits the user to do parallel configurations on selected cars. The user can very simply compare two different configurations, which differ in the color of the body. This comparison is done directly on the same screen by cutting the car image into two parts. With each of the two image parts being a different color, the user can choose options to continue the configuration process.

Therefore, the user has the opportunity to compare different product options throughout the course of the configuration and not only at the end, as happens on many other sites (such as those that have already been analyzed). The user can immediately discard the combinations of attributes that do not reflect his or her tastes and does not waste time completing separate configurations.

9.8.3.2 *Timbuk2*

Another interesting solution for the comparison of different product alternatives is adopted on the Timbuk2 website (www.timbuk2.com), where the user can choose different products simultaneously and compare them with each other on a single page after selecting the product features that interest him or her the most and configuring more product variants. The comparison is made based on the technical characteristics and features that differentiate each variant.

Furthermore, reviews from customers who have already bought and used the company's products are available. In these customer reviews, the user can find answers about the quality and reliability of these products, if he or she does not have them already, and can clarify indecisions that may emerge during the configuration process.

9.9 Conclusions

The analysis of web-based shoe configurators showed that, on average, shoe configurators are not inferior to other fashion product configurators. However, a number of them could improve some specific capabilities. In general, the capabilities that offer the greatest degree for improvement are *easy comparison* and, to a lesser extent, *benefit–cost communication* and *user-friendly product space description*. The specific examples of best practices that have been identified in this research can help improve shoe configurators and ultimately increase a potential customer's purchase intention. For the less advanced capabilities, examples of best practices from WBSCs of both fashion and nonfashion products are provided. These improvements in WBSC capabilities could contribute to enhancing self-expressiveness and uniqueness benefits, which currently obtain the lowest score among the benefits perceived by customers as a result of their shoe configuration experiences.

Acknowledgments

We acknowledge the financial support of (a) the Italian Ministry of Education, Universities, and Research under the Italian flagship program La Fabbrica del Futuro, project MADE4FOOT (FdF-SP2-T1.1) and (b) the University of Padova, project ID CPDA129273.

References

Addis, M., and Holbrook, M. B. (2001). On the conceptual link between mass customisation and experiential consumption: An explosion of subjectivity. *Journal of Consumer Behaviour*, 1(1), 50–66.

Aichner, T., and Coletti, P. (2013). Customers' online shopping preferences in mass customization. *Journal of Direct, Data and Digital Marketing Practice*, 15(1), 20–35.

Belk, R. W. (1988). Possessions and the extended self. *Journal of Consumer Research*, 15(2), 139–168.

Brewer, M. B. (1991). The social self: On being the same and different at the same time. *Personality and Social Psychology Bulletin*, 17(5), 475–482.

Fogliatto, F. S., da Silveira, G. J., and Borenstein, D. (2012). The mass customization decade: An updated review of the literature. *International Journal of Production Economics*, 138(1), 14–25.

Forza, C., and Salvador, F. (2008). Application support to product variety management. *International Journal of Production Research*, 46(3), 817–836.

Franke, N., Schreier, M., and Kaiser, U. (2010). The "I designed it myself" effect in mass customization. *Management Science*, 56(1), 125–140.

Grosso, C., Trentin, A., and Forza, C. (2014). Towards an understanding of how the capabilities deployed by a web-based sales configurator can increase the benefits of possessing a mass-customized product. In A. Felfernig, C. Forza, and A. Haag, Eds., *Proceedings of the 16th International Configuration Workshop*, September 25–26, 2014, Novi Sad, Serbia. *CEUR-WS* 1220, 81–88.

Heiskala, M., Tiihonen, J., Paloheimo, K.-S., and Soininen, T. (2007). Mass customization with configurable products and configurators: A review of benefits and challenges. In T. Blecker and G. Friedrich, Eds., *Mass Customization Information Systems in Business*, pp. 1–32. Hershey, PA: IGI Global.

McCarthy, I. P. (2004). Special issue editorial: The what, why and how of mass customization. *Production Planning and Control*, 15(4), 347–351.

Merle, A., Chandon, J.-L., Roux, E., and Alizon, F. (2010). Perceived value of the mass-customized product and mass customization experience for individual consumers. *Production and Operations Management*, 19(5), 503–514.

Pine II, B. J. (1993). *Mass Customization: The New Frontier in Business Competition*. Boston, MA: Harvard Business School Press.

Schreier, M. (2006). The value increment of mass-customized products: An empirical assessment. *Journal of Consumer Behaviour*, 5(4), 317–327.

Trentin, A., Perin, E., and Forza, C. (2011). Overcoming the customization-responsiveness squeeze by using product configurators: Beyond anecdotal evidence. *Computers in Industry*, 62(3), 260–268.

Trentin, A., Perin, E., and Forza, C. (2012). Product configurator impact on product quality. *International Journal of Production Economics*, 135(2), 850–859.

Trentin, A., Perin, E., and Forza, C. (2013). Sales configurator capabilities to avoid the product variety paradox: Construct development and validation. *Computers in Industry*, 64(4), 436–447.

Trentin, A., Perin, E., and Forza, C. (2014). Increasing the consumer-perceived benefits of a mass-customization experience through sales-configurator capabilities. *Computers in Industry*, 65(4), 693–705.

Tu, Q., Vonderembse, M. A., and Ragu-Nathan, T. S. (2001). The impact of time-based manufacturing practices on mass customization and value to customer. *Journal of Operations Management*, 19(2), 201–217.

chapter ten

Empirical investigation on implications of configurator applications for mass customization

Linda L. Zhang and Petri T. Helo

Contents

ABSTRACT

In pursuing mass customization, many companies have applied product configurators for configuring the right amount of product variety. While studies have been reported to shed light on how product configurator applications achieve time reduction and quality improvement in fulfilling mass customization, investigations addressing the implications of product configurator applications for companies' business activities have not been seen. However, understanding the implications is very important for companies because with such understanding, they can better plan actions and make changes to embrace the application of product configurators, thus further reaping the benefits of mass customization. Based on a survey, this study investigates the implications of product configurator applications. The results indicate (1) how product configurator applications affect companies' business activities, (2) the difficulties in designing, developing, and using product configurators, and (3) the potential barriers preventing companies from effectively applying product configurators in the future. The results show several improvement areas for companies to investigate in the future so as to achieve, to the largest extent, the benefits of implementing product configurators.

10.1 Introduction

Since the publication of Pine's book in 1993, many companies have adopted mass customization as a new manufacturing strategy to replace mass production, in hope of delivering diverse customized products while utilizing the available manufacturing capabilities (Zhang, 2007). In pursuing mass customization, companies have adopted product configurators, which are information systems for configuring the right amount of product variety (also called *customer-wanted variety* in the literature) (Zhang, 2014). In view of the benefits that they bring to companies, such as better-managed product variety, shortened sales delivery processes, and simplified order acquisition and fulfillment activities (Aldanondo et al., 2000; Forza and Salvador, 2002; Haug et al., 2011), product configurators have been receiving continuous interest from both academia and industry alike.

As a result of these countless efforts, numerous articles have been published to present solutions to diverse configurator-related issues. Many of these articles address configuration knowledge representation and modeling, configuration modeling and solving, and methods and approaches for product configurator design (e.g., Chen and Wang, 2009; Chu et al., 2005; Falkner et al., 2011; Felfernig, 2007; Felfernig et al., 2001; Ong et al., 2006). The feature common to these articles is that the solutions are proposed from a theoretical point of view and are validated using ad hoc industrial examples. In comparison, a relatively small number of articles deal with the practical issues related to product configurator applications. Among these articles, some empirically investigate how product configurators achieve lead time reduction and quality improvement (Haug et al., 2011; Trentin et al., 2011, 2012). Some use single cases to show (1) how product configurator development can be facilitated (Haug et al., 2010; Hvam et al., 2003; Hvam and Ladeby, 2007), (2) how product configurators contribute to variety management (Forza and Salvador, 2002), and (3) suitable product configurator development strategies (Haug et al., 2012).

Despite these efforts, to the best of our knowledge, limited investigations have been conducted to study the implications of product configurator applications for companies' business activities. Due to the lack of studies, several important issues related to configurator applications are unclear. How do product configurator applications affect companies' business activities? What are the objectives of implementing product configurators? Are the configurators designed, developed, and used in line with the objectives? What are the difficulties in designing, developing, and using product configurators in practice? What are the barriers potentially preventing companies from effectively applying configurators in the future? It is very important to have a clear understanding of all such issues. This is because such an understanding can help companies better design, develop, and apply product configurators, which in turn can help reap the benefits of product configurator applications to the largest extent.

In view of the lack of investigations and the unclear practical issues pertaining to product configurator applications, this study empirically addresses their implications for companies' business activities. Among all the aforementioned issues, it investigates (1) how product configurator applications affect companies' business activities, (2) the difficulties in designing, developing, and using product configurators, and (3) the potential barriers influencing the effective application of product configurators in the future. A survey is used to collect data for clarifying these three issues. With respect to how product configurator applications affect companies' business activities, the data results and analysis show the major tasks and users of product configurators, the functional units reorganized corresponding to the adoption of product configurators, the changes to business processes, to companies' legacy systems, and to the number of employees, and performance improvements thanks to the adoption of product configurators. Besides the aforementioned difficulties, the potential barriers influencing the effective application of product configurators include two new ones: unclear customer requirements and the unsafe feelings of employees. Based on the results, this study further highlights three areas for companies to investigate in the future, in hope of realizing the optimal benefits from implementing product configurators by carrying out suitable decisions and activities. They include IT capacity and capability enhancement, organizational redesign, and top-down support and company-wide engagement.

This chapter is organized as follows. The data collection and analysis methods are presented in Section 10.2. Results and corresponding analysis are provided in Section 10.3. Building on the insights in Section 10.3, we further provide in Section 10.4 the possible changes and improvements that companies may undertake for improving product configurator applications. Section 10.5 concludes the chapter by discussing limitations and potential future research.

10.2 Research methods: Data collection and analysis

As the purpose of this study is to investigate the implications of product configurator applications for companies' business activities, an empirical study was performed. In doing so, a questionnaire was developed for collecting data. The collected data was computed and analyzed to reveal the application implications and to identify the related opportunities as well.

In designing the questionnaire, we included questions related to matters such as the functions that product configurators perform, the business process and IT system changes caused by product configurator applications, the difficulties in implementing product configurator projects, the potential barriers preventing the effective application of product configurators in the future, and the performance improvements resulting from product configurator applications. Besides these questions, there are also general demographic questions such as position titles, company size, and industry type. (See Appendix 10.1 for the main questions used in the survey.) By considering the explorative nature of this study, nominal scales were used to present alternative choices for each question. Each respondent could select more than one alternative for each question based on his or her own product configurator application practice. In this regard, the respondents are not mutually exclusive with respect to alternatives in each question. The alternatives were determined based on the literature and our experiences of working with companies. To avoid missing potential alternatives, an "Other" option was included in each question for respondents to give details in addition to selecting the given alternatives.

To verify the initial questionnaire with respect to the sufficiency and appropriateness of questions, we pretested it in five Finnish companies with which we have collaborations. In the pretest, five company representatives completed the questionnaire and provided comments. In addition, we made phone contact for clarification of the comments and additional remarks. Based on the feedback, some questions were revised, leading to the finalized questionnaire.

Based on the finalized questionnaire, the survey was conducted in collaboration with EMpanel Online consulting company. After preselection, the questionnaire was sent to 305 companies, which had balanced distributions with respect to the company size, industry, and time duration (in year) of product configurator applications. We received 61 completed questionnaires and thus the response rate was 20%. The respondents were mainly IT managers or managers with sales IT responsibilities, and the companies represented the computer, telecommunication systems, and industrial machinery industries.

In analyzing the data, we computed the total occurrence of each alternative that was selected by the respondents and the corresponding percentage. In this regard, we analyzed the distribution of the selected alternatives for each question. As the respondents selected more than one alternative for most of the questions, each respondent was counted more than once in the computation of the percentage of selected alternatives for each question.

10.3 Results and analysis

In accordance with the questionnaire, the collected data was analyzed with respect to (1) how the application of product configurators affects companies' business activities, (2) the difficulties in designing, developing, and using product configurators, and (3) the barriers potentially influencing the effective application of product configurators in the future.

10.3.1 Product configurator applications affecting companies' business activities

In studying how companies' business activities are affected by the application of product configurators, it is essential to understand the major tasks that product configurators perform. This is because these tasks contribute to companies' activities in designing, producing, and delivering products. It is equally important to understand the major users of product configurators. In accordance with configurators' tasks and users, there might be changes to companies' business processes, functional units, IT systems, employee numbers, and operational performance.

10.3.1.1 Major tasks of product configurators

The literature suggests, either directly or indirectly, that a product configurator carries out diverse tasks. The survey results confirm this. As shown in Figure 10.1, the tasks that configurators perform can be classified into three groups: sales order processing, product documentation, and production documentation. Configurators in 62% of the respondents perform sales order processing, which includes quotation preparation,

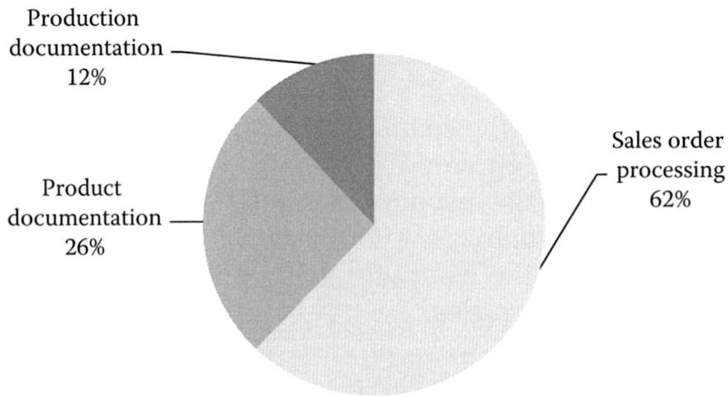

Figure 10.1 Major tasks of product configurators and corresponding percentages.

sales order specification, and product specification. While configurators in 26% of the respondents carry out product documentation, such as bill of materials (BOMs) and drawing generation, in 11% of the respondents, they perform production documentation, such as routing generation and process plan generation. (Note that the word *respondent* means the responding company from this section onward.) While published articles provide anecdotal evidence, there is no study presenting such a complete distribution of the major tasks that product configurators perform. Based on our analysis method (Section 10.2), the respondents are not mutually exclusive in these percentages. This is the same for the rest of the results.

10.3.1.2 Main users of product configurators

The main users of product configurators include end customers, sales staff, production planners, and product designers, as shown in Figure 10.2. Also provided in Figure 10.2 are the percentages of respondents that selected

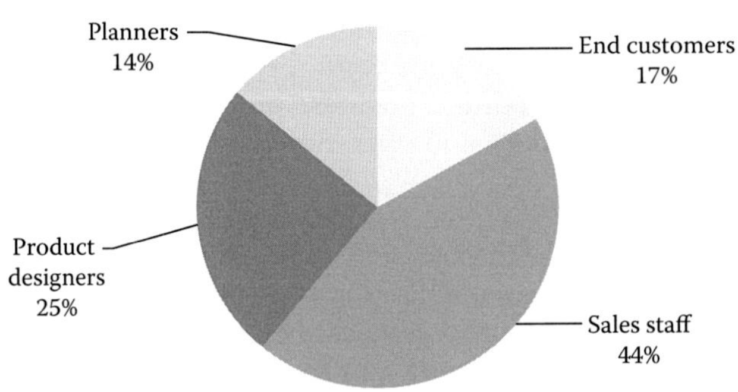

Figure 10.2 Main users of product configurators and corresponding percentages.

the corresponding users. Sale staff include both sales agents and internal sales staff; product designers include product designers and product developers. Being consistent with the major tasks that configurators perform, product configurators are used by end customers in 17% of the respondents, sales staff in 44%, designers in 25%, and production planners in 14%. Similarly, there are no studies presenting such a complete distribution of the main users of product configurators.

10.3.1.3 Functional units reorganized

Figure 10.3 shows the results with respect to reorganized functional units caused by product configurator applications. The percentages in the figure represent the percentages of respondents that selected each of the given alternatives, including (1) no change made to functional units, (2) sales unit is affected and reorganized, (3) product design and development unit is affected and reorganized, and (4) production unit is affected and reorganized. As product configurators take over tasks that were performed previously by different functional units, their applications may bring many changes to the organization of these units. According to the survey result, both the sales and product design units in 33% of the respondents are affected and thus reorganized, and the production unit in 18% of the respondents is reorganized (Figure 10.3). It is interesting to see that product configurator applications do not bring any changes to the functional units in 16% of the respondents. Our experiences of working with a number of companies show that this finding is rational. This is especially true in situations where the end customers are the main users of product configurators. In this regard, 16% is not a surprising figure considering the result: end customers are the users in 17% of the respondents. However, this connection needs to be verified in future studies.

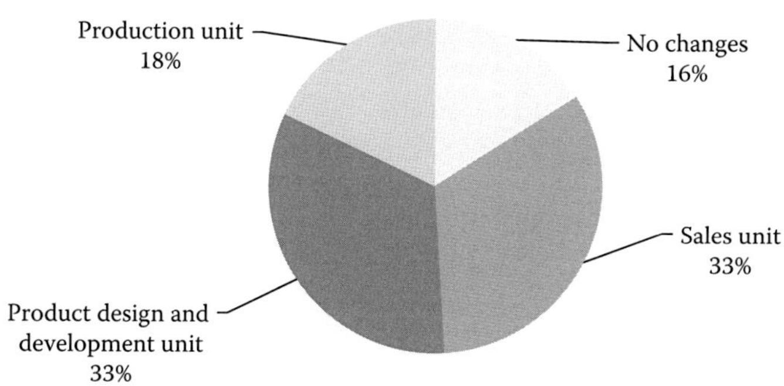

Figure 10.3 Function units reorganized and corresponding percentages.

10.3.1.4 Business process changes

As product configurators automatically perform many activities that were carried out manually in the past, their applications might incur business process changes as well. Our survey result confirms this. In the survey, many respondents indicate that their business processes have been changed. These changes include the following: (1) the original manual quotation preparation is done automatically by configurators in 24% of the respondents, (2) specifying product functions is done automatically by configurators, (3) product technical details (e.g., BOMs and drawings) are generated automatically by configurators, and (4) manufacturing documents are generated automatically. The percentages of respondents where these changes occur are 24%, 27%, 21%, and 15%, respectively, as shown in Figure 10.4. While most respondents experience business process changes, 13% of the respondents indicate that product configurator applications do not bring changes to their business processes (see Figure 10.4). This seems consistent with what we have found: 16% of respondents point out that product configurator applications do not cause changes to the functional units. When there are no changes to companies' functional units, there may not be changes to the business processes.

10.3.1.5 Changes to companies' legacy systems

In performing tasks, product configurators interact with companies' other IT systems for receiving inputs and/or sending outputs. Consequently, product configurator applications may cause changes to companies' legacy systems. The results in Figure 10.5 confirm this. As shown, the following systems are modified to link with product configurators: (1) design systems, (2) production planning and control systems, (3) material requirements planning systems, and (4) accounting systems. The percentages of

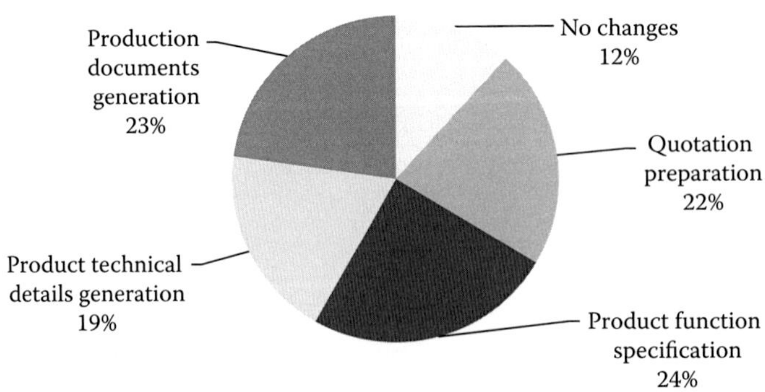

Figure 10.4 Business process changes and corresponding percentages.

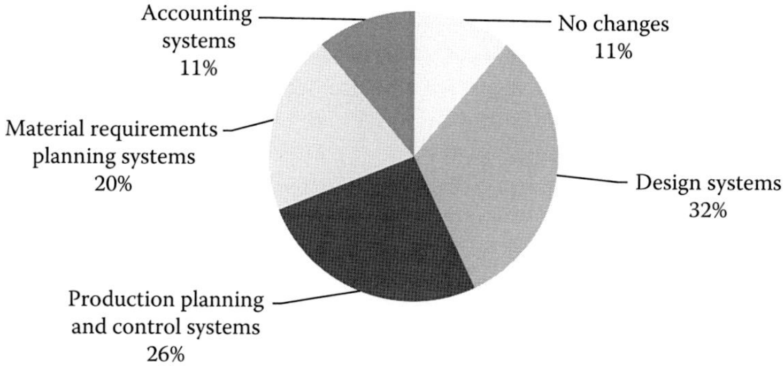

Figure 10.5 Changes to legacy systems and corresponding percentages.

the respondents that experience each of these changes are 32%, 26%, 20%, and 11%, respectively, as shown in Figure 10.5. Lastly, 11% of the respondents indicate that there are no changes to their existing legacy systems (Figure 10.5). When product configurators are built-in modules of enterprise resource planning systems, it might be possible that companies do not need to modify their legacy systems. Further studies are needed to investigate this issue.

10.3.1.6 Changes to the numbers of employees

As product configurators automatically perform many activities that were performed manually in the past, intuitively, product configurator applications should reduce the number of full-time employees. However, it is surprising to see that 67% of the respondents hired full-time employees, whereas only 5% indicate that they laid off employees, as shown in Figure 10.6. A share of 28% of the respondents claim that there is no change to their number of employees. As indicated in the following results (Sections 10.3.2 and 10.3.3), companies do not have sufficiently skilled IT system designers and developers. In this regard, the application of product configurators may lead to the recruitment of new employees.

10.3.1.7 Performance improvements

The available literature anecdotally reports diverse performance improvements resulting from the application of product configurators. This study finds similar results (Table 10.1), thus supporting the literature. In Table 10.1, the first column indicates the performance measures considered, whereas the first row provides the alternative improvement ranges (in percentages). The values in the cells present the percentages of respondents that achieved the improvements falling into the corresponding

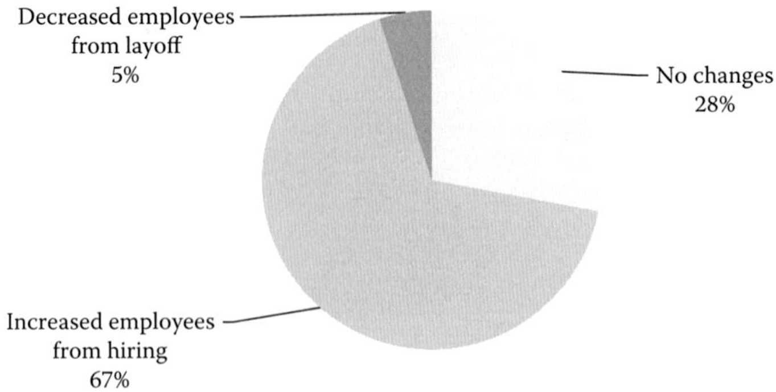

Figure 10.6 Changes to the number of employees and corresponding percentages.

ranges for each performance measure. As shown, for all the respondents, the improvements include (1) increased sales volume, (2) increased correct sales orders, (3) reduced production rework, (4) increased customer orders accepted, (5) reduced order-processing time, and (6) reduced sales delivery time.

As shown in Table 10.1, for four performance measures, including increased sales volume, increased correct sales orders, reduced production rework, and increased customer orders accepted, most respondents achieve an improvement ranging from 30% to 50%, as indicated by their corresponding percentages: 43%, 46%, 37%, and 31%. As seen from the table, it is difficult for most respondents to achieve higher improvements of these measures. For example, only 6% of respondents increased their sales volume higher than 80%. For reduced order-processing time and reduced sales delivery lead time, the majority—28% and 33% of the respondents, respectively—achieve improvements by no higher than 30%. As there are more interactions among different functional units in the

Table 10.1 Performance improvements achieved by the respondents

	Alternative improvement ranges			
Performance measures	0–30	30–50	50–80	Higher than 80
Increased sales volume	18	43	33	6
Increased correct sales orders	8	46	27	19
Reduced production rework	30	37	25	8
Increased customer orders accepted	20	31	31	18
Reduced order-processing time	28	25	25	22
Reduced sales delivery lead time	33	26	21	20

Note: All values are in percentages.

entire cycle of delivering sales orders than in order processing, improvements in sales delivery lead time might be lower than in order-processing time. However, it is interesting to see that respondents achieve similar improvements in order-processing time and sales delivery lead time. Further investigations should be carried out to dig deeper into this issue.

10.3.2 Product configurator design, development, and usage difficulties

The results show that most respondents experience difficulties in designing, developing, and using product configurators, and 50% of the respondents indicate that it is rather difficult for them to design product configurators. The two major reasons are (1) a lack of IT system designers (in 50% of the respondents) and (2) a problem where IT system designers and product designers cannot communicate well (in 45% of the respondents). With our experience of working with companies and based on the literature, these results are understandable. Manufacturing companies normally hire IT engineers for maintaining systems in support of their core business activities: design and production. In this regard, the IT engineers may not possess sufficient skills and capabilities for designing product configurators. The early literature points out that due to the differences in communication languages, configurator designers and product experts have difficulties in communicating effectively (Haug et al., 2010). The results found in this study are consistent with the literature.

Similarly, most respondents have difficulties in developing product configurators. The biggest challenge for companies to develop product configurators is their high complexity, as supported by 52% of the respondents. The other two main difficulties include (1) the lack of good IT system developers (in 24% of the respondents) and (2) the continuous evolution of products and the resulting high product complexity (in 24% of the respondents). While product complexities do not appear as a main difficulty in designing configurators, they do cause difficulties in developing product configurators. This is because, in accordance with product complexities, the product configurator design is complex too. It is understandable that complex product configurators are difficult to develop.

In using product configurators, companies also have difficulties. These are caused by (1) nonuser-friendly interfaces (in 44% of the respondents), (2) inefficient communications for getting the required inputs (in 31% of the respondents), (3) the high complexity of product configurators (in 12.5% of the respondents), and (4) a lack of sufficient training (in 12.5% of the respondents). In processing customer orders, configuring products, and generating product/manufacturing documents, configurators require diverse inputs. These inputs originate from customers, sales staff, designers, and so on. In many cases, the input providers are from different offices or even different companies. This location dispersion may hinder the effective communications for required inputs. Additionally, even in situations where the input providers are in the same location, due to, for example, other tasks that they need to deal with, the input providers may not be able to supply the required inputs on time. In our view, the other three difficulties are all interconnected with one another. Firstly, complex product configurators may have many interrelated modules and procedures. Secondly, more training is required to understand and use these modules and procedures. However, companies are busy dealing with daily operations activities and may not provide enough training time to users. As shown in practice (e.g., the fail of SAP project in Avon (http://blogs.wsj.com/cio/2013/12/11/avons-failed-sap-implementation-reflects-rise-of-usability), caused by design difficulties, complex IT systems tend to have nonuser-friendly interfaces. In this regard, the nonuser-friendly interfaces may be, at least, partially related to product configurator complexities.

10.3.3 *Potential barriers influencing product configurators' effective application*

One question in the questionnaire asked respondents for the barriers that may potentially prevent them from effectively applying product configurators in the future. The results here are consistent with those discussed in Section 10.3.2. The earlier results show that companies have difficulties in designing and developing product configurators because of the lack of technical IT staff. Similarly, the lack of IT staff also appears to be a major barrier for companies to effectively use product configurators in the future. The fact that products keep evolving is one of the three difficulties in using product configurators is also acknowledged as one barrier for future application. Two additional barriers, unclear customer requirements and the unsafe feelings of employees, are brought up. Due to the insufficient knowledge about the products, customer requirements are often imprecise and ambiguous (Jiao and Zhang, 2005). In addition, they often conflict with one another or are subject to change (Kristianto et al., 2015). As product configurators need articulated customer requirements, ambiguous and conflicting requirements will negatively affect the effective application of product configurators. As a matter of fact, this was indicated by some of the five Finnish company representatives during the initial questionnaire pretest. As product configurators execute activities that were previously carried out by employees, the affected employees perceive product configurators as a menace to their positions (Forza and Salvador, 2002). In this regard, the unsafe feelings of employees may become an obstacle for the effective application of product configurators in the future.

In summary, the largest barrier is the continuous evolution of products, as pointed out by 75% of the respondents. Other barriers include (1) a lack of technical IT staff for maintaining the configurators (seen by 47% of the respondents), (2) unclear customer requirements (perceived by 47% of the respondents), and (3) the unsafe feelings of employees because of the possibility of losing jobs (agreed by 34% of the respondents).

10.4 Discussions

Along with the benefits achieved, product configurator applications bring many additional requirements and changes to companies' existing ways of doing business, as shown in this study. While the changes and requirements may not be perceived as beneficial, they open up opportunities for companies to find new ways of doing business that involve product configurators. Based on data analysis and results, this study highlights three areas for investigation: (1) IT capacity and capability enhancement, (2) organization redesign, and (3) top-down support and company-wide engagement.

10.4.1 IT capacity and capability enhancement

Product configurators are basically IT systems. The optimal design and development of these systems will bring many advantages to companies, such as the configuration of optimal products, the cutting down of configuration time, the reduction of configuration errors, easy application, the reduction of training time, and so on (Haug et al., 2012). Such design and development demand system designers and developers with sufficient skills and experience. However, the results indicate that many companies do not have them. It will be beneficial to companies, especially those that design and develop configurators in house, to have sufficient well-trained system designers and developers. These designers and developers bring companies additional IT capacities and capabilities. Developing such IT capacities and capabilities can also be justified by other issues. The fact that products keep evolving necessitates continuous maintenance and upgrading (Section 10.3.2). Due to their complexity, configurator maintenance and upgrading are not easy tasks to perform. In addition, if they are not performed on time, companies may delay product configuration, production, and delivery. This may, in turn, cause companies to lose customers. In this regard, sufficient well-trained system designers and developers can also contribute to the continuous maintenance and upgrading of configurators and product models.

10.4.2 Organizational redesign

Product configurator applications bring many changes to companies' existing activities, processes, and functional units (Section 10.3.1). While simply reorganizing the affected units, as the practice does (see Section 10.3.1.3, "Functional units reorganized"), may to a certain degree facilitate product configurator applications, it is insufficient for companies to realize the full benefits of product configurators (Salvador and Forza, 2004). In fact, the communication difficulties (Section 10.3.2) lend themselves to this point. In accordance with the tasks and functions that product configurators perform, companies should reorganize their business processes and structures by reallocating the responsibilities of each individual employee and functional unit. The reorganization should be performed such that each employee has a clear vision of his or her activities, tasks, and responsibilities. The same applies for the functional units. Besides, information exchange protocol and procedures need to be (re)designed such that communication difficulties in applying product configurators can be eliminated. Lastly, as one of the potential barriers for effective configurator application in the future lies in unclear customer requirements, some efforts in organization redesign may be directed at suitable tools, techniques, systems, and so on, and the related issues, for obtaining clear customer requirements or properly changing the management process.

10.4.3 Top-down support and company-wide engagement

As with the implementation of any new technologies, the implementation of product configurators needs continuous support and commitment from all levels in a company, especially the top management level. This support and commitment is very important for completing the necessary organizational changes and for successfully implementing product configurator projects. The literature shows that a lack of long-term commitment is one of the main reasons for the failure of many technology implementation projects (Bergey et al., 1999). As employees, including the middle-management level, have a tendency to resist changes (Paper and Chang, 2005), regular encouragement and incentives from the top management level are required to remove employees' hostile attitudes toward the application of product configurators. Once the employees look positively at product configurator applications, they are willing to accept and implement organizational changes. As perceived by companies, employees' unsafe feelings over losing their jobs due to process automation is one of the important barriers potentially preventing companies from effectively applying configurators in the future (Section 10.3.3). To encourage employees and remove their unsafe feelings, the top management level should create more training activities. With these training activities, the employees may master additional skills. They may also involve employees in the company's important meetings, share with employees the company's daily or weekly news and developments, and so on. All these supports may help employees to regain their confidence and develop correct attitudes toward configurator applications.

10.5 Conclusion

In view of the contribution of product configurators to pursuing mass customization, this study investigated the implications of product configurator applications for companies' business activities. The belief is that it is beneficial for companies to understand the difficulties and challenges before embarking on a product configurator project. As shown in the results, product configurator applications bring many changes and difficulties along with performance improvements. The changes together with the difficulties highlight a number of areas to be investigated if companies want to achieve the optimal benefits from using product configurators. These improvement areas include (1) the development of IT capacity and capability for addressing product configurator design, development, and maintenance, (2) the reorganization of the company's structure (e.g., individuals' and functional units' responsibilities and tasks, information exchange protocol) for eliminating communication difficulties in configurator design, development, and application, and (3) top-level support and company-wide engagement, which is fundamental to the achievement of the previous two.

In conducting the survey, we used nominal scales by considering the explorative nature of this study. While the nominal scale permits an easily understandable questionnaire, it makes analysis less exact than a Likert scale. In other words, it may not be able to identify the causal relationships among the interesting elements involved in the implications of configurator applications. In this regard, this study highlights an interesting future research topic. An extended quantitative method involving data analysis based on Likert scales might be conducted to reveal these causal relationships. Moreover, since product configurator applications incur business process changes and require organizational redesign, business process reengineering, where product configurators are applied, might deserve future research efforts as well.

References

Aldanondo, M., Rouge, S., and Reron, M. (2000). Expert configurator for concurrent engineering: Cameleon software and model. *Journal of Intelligent Manufacturing*, 11, 127–134.

Bergey, J., Smith, D., Tiley, S., Weiderman, N., and Woods, S. (1999). Why reengineering projects fail. *Carnegie Mellon Software Engineering Institute: Product Line Practice Initiative*, 1, 1–30.

Chen, Z., and Wang, L. (2009). Adaptable product configuration system based on neural network. *International Journal of Production Research*, 47(18), 5037–5066.

Chu, C. H., Lin, C. W., Li, Y. W., and Yang, J. Y. (2005). Online product configuration in e-commerce with 3D viewing technology. *International Journal of Electronic Business Management*, 3(3), 225–234.

Falkner, A., Haselbock, A., Schenner, G., and Schreiner, H. (2011). Modeling and solving technical product configuration problems. *Artificial Intelligence for Engineering Design, Analysis and Manufacturing*, 25, 115–129.

Felfernig, A. (2007). Standardized configuration knowledge representations as technological foundation for mass customization. *IEEE Transactions on Engineering Management*, 54(1), 41–56.

Felfernig, A., Friedrich, G., and Jannach, D. (2001). Conceptual modeling for configuration of mass-customizable products. *Artificial Intelligence in Engineering*, 15, 165–176.

Forza, C., and Salvador, F. (2002). Managing for variety in the order acquisition and fulfillment process: The contribution of product configuration systems. *International Journal of Production Economics*, 76, 87–98.

Haug, A., Hvam, L., and Mortensen, N. H. (2010). A layout technique for class diagrams to be used in product configuration projects. *Computers in Industry*, 61, 409–418.

Haug, A., Hvam, L., and Mortensen, N. H. (2011). The impact of product configurators on lead times in engineering-oriented companies. *Artificial Intelligence for Engineering Design, Analysis and Manufacturing*, 25, 197–206.

Haug, A., Hvam, L., and Mortensen, N. H. (2012). Definition and evaluation of product configurator development strategies. *Computers in Industry*, 63, 471–481.

Hvam, L., and Ladeby, K. (2007). An approach for the development of visual configuration systems. *Computers and Industrial Engineering*, 53, 401–419.

Hvam, L., Riis, J., and Hansen, B. L. (2003). CRC cards for product modeling. *Computers in Industry*, 50, 57–70.

Jiao, J., and Zhang, Y. (2005). Product portfolio identification based on association rule mining. *Computer-Aided Design*, 37(2), 149–172.

Kristianto, Y., Helo, P., and Jiao, R. J. (2015). A system level product configurator for engineer-to-order supply chains. *Computers in Industry*, 72, 82–91.

Ong, S. K., Lin, Q., and Nee, A. Y. C. (2006). Web-based configuration design system for product customization. *International Journal of Production Research*, 44(2), 351–382.

Paper, D., and Chang, R. (2005). The state of business process reengineering: A search for success factors. *Total Quality Management*, 16(1), 121–133.

Pine, B. J. (1993). *Mass Customization: The New Frontier in Business Competition*. Boston, MA: Harvard Business School Press.

Rosenbush, S. (2013). Avon's failed SAP implementation reflects rise of usability. *Wall Street Journal* December 11. http://blogs.wsj.com/cio/2013/12/11/avons-failed-sap-implementation-reflects-rise-of-usability/.

Salvador, F., and Forza, C. (2004). Configuring products to address the customization-responsiveness squeeze: A survey of management issues and opportunities, *International Journal of Production Economics*, 91(3), 273–291.

Trentin, A., Perin, E., and Forza, C. (2011). Overcoming the customization-responsiveness squeeze by using product configurators: Beyond anecdotal evidence. *Computers in Industry*, 62, 260–268.

Trentin, A., Perin, E., and Forza, C. (2012). Product configurator impact on product quality. *International Journal of Production Economics*, 135, 850–859.

Zhang, L. (2007). Process platform-based production configuration for mass customization. PhD dissertation, Division of Systems and Engineering Management, Nanyang Technological University, Singapore.

Zhang, L. (2014). Product configuration: A review of the state-of-the-art and future research. *International Journal of Production Research*, 52(21), 6381–6398.

Appendix 10.1

Questions in the survey

General demographic questions:

1. Are you knowledgeable about the design or development or application of the product configurator?
2. What is your position title?
3. What is your industry?
4. What is your company size?
4. How long has the configurator been in use?
5. How many types of products is the configurator used for?

Product configurator applications:

1. Which tasks are performed by the configurator?
2. Who are the users?
3. Which functional units were reorganized?
4. Which business processes were changed?
5. Which changes were made to other computer systems?
6. Did you layoff/hire full-time employees?

Performance of product configurator applications:

1. The percentage of increased sales volume.
2. The percentage of increased correct sales orders.
3. The percentage of reduced production rework.
4. The percentage of increased customer orders accepted.
5. The percentage of reduced order-processing times.
6. The percentage of reduced sales delivery times.

Difficulties in implementing the product configurator project:

1. The difficulties in designing the product configurator.
2. The difficulties in developing/maintaining the configurator.
3. The difficulties in using the product configurator.

Barriers preventing companies from continuously applying the product configurator:

1. The continuous evolution of products.
2. The lack of technical staff for maintaining the configurator.
3. Customer requirements are rather unclear.
4. The unsafe feelings of employees because of the possibility of losing jobs.

chapter eleven

Sustainability assessments for mass customization supply chains

Carlo Brondi, Davide Collatina, and Rosanna Fornasiero

Contents

ABSTRACT

The evaluation of the environmental impact of production networks has been under debate during the last few years. Currently, there is a shift of production paradigm from mass production to customization and personalization.

The aim of this chapter is to evaluate the sustainability of supply chains, applying a model based on the integration of life cycle assessment (LCA) with discrete simulation to compare different customization policies in a networked context. In the developed model, the environmental impact of the supply chain is assessed through an innovative modular LCA where different levels of customization have been analyzed. The chapter then compares the scenarios based on variations of drivers such as lead time to the customer, quality in terms of scraps, and the level of sustainability of the suppliers. The model is validated by collecting data from a fashion-based case study taking into consideration the environmental impact of a certain batch production. The preliminary results highlight that specific decisional areas under the control of supply managers (e.g., supplier selection and manufacturing defects) can significantly affect the environmental impact of the whole supply chain.

11.1 Introduction

The quantitative assessment of environmental sustainability is a recurrent area of interest for evaluating production phases, transportation, and suppliers within the literature. Sustainability assessments also concern modern production paradigms such as knowledge-intensive services to customers (Gallouj et al., 2015). In particular, Petersen et al. (2011) address the issue whether the concepts of mass customization and sustainability are fundamentally compatible. The updated mass customization paradigm calls for both personalized outputs and cost/eco-efficiency tracking in order for companies to maintain their competitiveness and create value (Mourtzis and Doukas, 2014; Ueda et al., 2009). The development of customized production and its related services seems implicitly to call for new collaborative supply chains (Romero et al., 2014) as well as for reliable models for sustainability characterization going beyond qualitative assessment (Kohtala, 2014). The preliminary involvement of consumers within product service systems, including referenced sustainability tracking, seems to also produce positive effects in fashion sectors (Armstrong et al., 2015). At the current stage, different studies have proposed alternatives for the design of sustainable supply chains (SSCs) and eco-efficient products in order to be compliant with the mass customization paradigm (Piplani et al., 2007; Lee and Huang, 2011; Govindan et al., 2014; Osorio et al., 2014), but the literature review emphasizes the lack of verification criteria in the presence of diverging possible effects due to customization policies (Kohtala, 2014). Positive environmental effects account for the reduction of preconsumer waste, lower transport emissions, minor product replacements, greater potential for remanufacturing, intermediary reduction, and use phase extension for customized products. Possible negative effects instead account for the augmented difficulties in product reuse, the need for energy/resource-intensive transformation processes, exposure to uncontrolled emissions in local environments, and the possible failure to replace traditional mass production.

From an industry perspective, despite widespread agreement on the importance of sustainability aspects for long-term competitive advantages, often companies need strong triggers in order to put into action initiatives for integrating these dimensions in their strategies. On one side, legal regulations, responsibility to stakeholders, customer demand, reputation loss, and environmental and social group pressures are often listed as triggers for companies to implement sustainability. On the other side, some barriers to implementing actions for an SSC are (Piplani et al., 2007)

1. The cost of implementing measurement systems for sustainability
2. Problems defining a value for the output with respect to environmental outcomes
3. The perception that data to be collected from different actors in the network is not manageable and will have a low impact on the global outcome
4. Difficulties taking unpopular and high-priced decisions for the network

In order to overcome these limits, researchers and managers are trying to answer the following questions:

- How should a supply chain accomplish the trade-off between economic and noneconomic objectives while making managerial decisions?
- What activities are necessary to implement an SSC?
- Which types of incentives are necessary to induce people to pursue sustainability objectives? (Noci, 1997)

When dealing with networked companies, the availability of data on time, quality, service, and so on of suppliers along the network is state of the art, while for environmental impact analyses there are still problems with the sharing of data that is considered confidential (e.g., energy consumption, water and heating release) and once the data is shared to make it homogeneous. Specific sectoral inventory data is useful to calculate the global warming potential (GWP) of a company and of a network.

11.2 State of the art

In the current business environment, the purchasing process has become a critical activity for adding value to products and a vital determinant to ensure the competitiveness of a company. This process becomes more complicated when environmental issues are considered because *green* purchasing must consider the supplier's environmental responsibility, depending on product chain assets, in addition to traditional factors such as the supplier's costs, quality, lead time, and flexibility. The management of suppliers based on strict environmental compliance seems to be not sufficient in view of a more proactive or strategic approach. Noci (1997) designed a green vendor–rating system for the assessment of a supplier's environmental performance based on four environmental categories—namely, green competencies, current environmental efficiency, green image, and net life cycle cost. The main limit in attributing a unique environmental performance index to a company seems to be linked to the management of a reliable, quantitative set of scientific values that can be considered constant in different comparisons. While literature related to supplier evaluation is plentiful, the works on green supplier evaluation or supplier evaluation that considers environmental factors are rather limited (Handfield et al., 2002; Humphreys et al., 2003). Two general aspects seem, then, to emerge as relevant in the sustainability assessment of mass customization: from one side it seems important to identify the sustainability features for a proper assessment, while on the other hand the environmental assessment of scalable product chains requires specific modeling issues. These aspects are faced separately in the next two sections.

11.2.1 Sustainable supply chains in mass customization

The high variability of customer demand and legislative pressure in EU countries on environmental aspects push academic and industrial communities to tackle the question of how to implement sustainable production systems. In order to accomplish this objective, a strong integration among the units of the supply chain is necessary and can help to maintain and build a durable competitive advantage with respect to competitors. For this reason, in the last few years many approaches have been proposed in international journals to support the implementation of SSCs (Dyllick and Hockerts, 2002; Seuring and Muller, 2008). The result of this academic and corporate interest has been the achievement of important goals for the sustainable success of firms in terms of integrated supply chains, green supply chains, the ecology industry, and long-term competitive advantages.

Despite there being, in recent years, widespread agreement on the importance of sustainability aspects for long-term competitive advantages, often companies need strong triggers in order to put into action initiatives for integrating these dimensions. Legal regulations, responsibility to stakeholders, customer demand, reputation loss, and pressure from environmental and social groups are often listed as triggers for companies to implement sustainability.

Zhu et al. (2008a) identify five *green supply chain management* (GSCM) factors: internal environmental management (IEM), green purchasing

(GP), cooperation with customers (CC) including environmental requirements, eco-design practices (ECO), and investment recovery (IR). Zhu et al. (2008b) present the implications in GSCM for closing the loop of the supply chain.

It is clear that the adoption of green practices impacts on environmental results—for example, in terms of pollution reduction (Klassen and Whybark, 1999)—but at the same time companies need to take over other environmental dimensions without forgetting to pursue profit objectives. In the literature, we can find some references to the positive role that environmental management plays in order to achieve operational performance (and it is established that operational performance is strictly and positively linked to financial performance), linking the *lean* and the *green* approach to management.

Hart (1997) and Florida (1996) suggest that environmental management can also provide cost savings, by increasing efficiency in production processes and improving the firm's performance, by facilitating the creation of resources and capabilities as well as the ability to innovate (Porter and Van der Linde, 1995; Russo and Fouts, 1997; Reinhardt, 1999). Moreover, Rusinko (2007) suggests a positive impact of pollution prevention on cost savings and competitive advantage. Christmann (2000) draws on the resource-based view of the firm and finds a moderating effect of innovation and implementation on the relationship between environmental practice and cost advantage.

On the other hand, the literature also raises a trade-off issue between environmental initiatives and operational performance (Clark, 1994; Walley and Whitehead, 1994), but in more recent works the impact of the cost of compliance with environmental goals was evaluated (Yu et al., 2009). For this reason, a "lean and green" perspective is adopted in the development of the performance measurement system, in order to monitor and control the trade-offs resulting from the implementation of environmental management.

The current debate on the customization paradigm poses a number of further issues for the sustainability paradigm. Customer-driven manufacturing could in fact address the reduction of environmental impact since the closest link between manufacturer and customer can imply a reduction of the environmental load due to operation and distribution (e.g., electricity, heating, and transport). The reduction of item stock and the increase in the value of traditional products (Bernard et al., 2011; Zhang et al., 2011; Bruno et al., 2013) seem to contribute to reducing environmental impact, particularly in product distribution to customers, in its use, and in the eventual recovery phase. Proper product modularization and proper efficiency policies in factory management can be the best way to increase efficiency as well as to counterbalance the negative effects of customization. Another open issue concerns transportation reduction, which depends on the supply chain configuration. The downsizing of the transport network could, in fact, conflict with the reduction of the efficiency of economies of scale. According to some authors, the relative environmental contributions from the stages of supply, manufacture, and waste production are affected by a strong sectoral characterization (Su et al., 2015).

Different authors have proposed simulation and optimization techniques to manage such divergent aspects. Mourtzis et al. (2013, 2014) proposed a toolbox to deal with the conflicting supply chain drivers in a network setup, metaheuristic and artificial intelligence methods, integrating assessment of carbon footprint limited to standard transport processes. A complete life cycle approach for supply chain carbon footprint modeling was proposed by Trappey et al. (2012) by using an I-O matrix in a real supply chain case based on three areas of investigation: materials, production, and logistics. Despite the completeness of the approach, assessment is developed in the presence of the same service model and by using sectoral data to assess the clustered carbon footprint. An effort to integrate LCA and business models with a mass customization perspective is made by Boër et al. (2013), where a complete reference set of environmental indicators for the modular subdivision of the whole product life cycle has been applied at the level of a single component, life cycle phase, and supplier. The detailed set of equations can, however, be difficult to implement in common multitier networks in which other life cycle inventory (LCI) indicators are available and the distinction between part manufacturing and the assembly phase is often not clear.

As a primary issue, then, a proper characterization of the real effects seems necessary (Su et al., 2015). In this respect, the environmental impact characterization due to current industrial practices is also affected by serious operative limits. In particular, the definition of a company's environmental performance is generally based uniquely on the type of transformation process or, instead, is referred to as standard operating conditions that are focused on a single product type. Real industrial practices are instead based on ever-changing production item batches.

11.2.2 *Environmental impact assessment of networks for customization strategies*

As mentioned in the introduction, it is important to underline that the customization process involves divergent environmental effects. On one hand, the simultaneous presence of these effects can lead to a higher impact from customization processes compared with processes for mass-produced items. Traditional processes can in fact benefit from scale economies. On the other hand, the comparison between a customized product and a standard product should require the same functional unit—that is, the performed service toward the final consumer. In this view, product customization is an additional service toward the consumer that changes the traditional functional unit of the mass customized products. According to these premises, environmental impact minimization can be relevant for the identification of the best implementation scenarios rather than a single comparison of a customized product against similar traditional products.

LCA represents the proper methodology to assess product environmental sustainability, and causes the intrinsic perspective on the whole product life cycle (Hugo and Pistikopolous, 2005; Bojarski et al., 2009; Nwe et al., 2010; Brondi et al., 2014). Nevertheless, the proper adoption of traditional LCA in the customized environment should overcome the following barriers:

- *Alignment between the life cycle perspective and the business perspective*: Inventory schemes for physical flows within small-to-medium enterprises can require a business-compliant approach that can significantly differ from LCI schemes. Internal operations can be committed to external suppliers so that mass and energy tracking is interrupted. Furthermore, the capability to provide reliable data from companies should not overcome the limited extent of the product life cycle (i.e., from first supply level up to final product distribution). Corporate environmental policies usually have to decide how many product chain levels should be included within the data inventory process. Ideally, the entire value chain should be analyzed, but resources and data availability can impose serious constraints on the assessment models (Brondi et al., 2012; ISO/TS, 2014; Unep, 2015). In a factory perspective, the *knowledge horizon* of the product manager can cover the background phases up to a certain supplier and the foreground phases up to the gate for the customer (Figure 11.1).

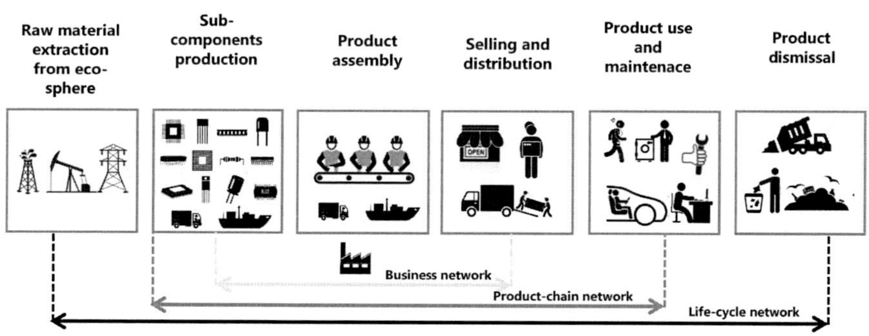

Figure 11.1 Knowledge horizons in modeling the environmental impact of the product chain.

- *Adaptation of inadequate data to new LCA studies* (Hagelaar and van der Vorst, 2002): Life cycle analysts can frequently abuse the literature and general-purpose databases in place of supply chain data in cases where the assessment involves a limited view on the product chain.
- *Misalignment between company environmental assessments and product design*: The designer and the life cycle analyst can require radically different procedures in order to modify the final solution (Brondi et al., 2012). Furthermore, the design of a product requires a set of parameters that can be insufficient for an environmental impact assessment.
- *Limited extent in the reuse of previous LCA studies* (Klöpffer, 2012): The literature suggests that a study review starts with a draft *goal and scope* chapter. In fact, each LCA is performed under specific assumptions and purposes; for example, an LCA for comparative assessment requires different rules from an LCA for internal assessment. Changes in the functional unit, in single processes, or in the system boundaries can then revoke the study validity for other purposes.
- *Uncertainty in the life cycle determination*: Existing product benchmarks commonly provide results with reference to the entire life cycle of a single product. The proper determination of life cycles requires the statistical tracking of a certain stock of products. Such stock involves different life cycles. The combined variance of specific environmental drivers is then fundamental. As an example, economy of scale, transport networks, stock variance, and environmental profiles from different suppliers can influence the variance analysis.
- *Misalignment between consequential and attributional methodologies*: LCA studies that aim to optimize a supply chain should compare different configurations of technologies and materials. The resulting comparative studies (consequential methodologies) require the assessment of additional marginal effects that can be difficult to model (e.g., marginal demand for a certain choice and avoided impacts). On the other hand, noncomparative studies (attributional methodologies) focus on the life cycle for a specific product. In particular, attributional methodologies make use of allocation factors requiring an impact subdivision according to eventual coproducts and services.

11.3 Proposal for an integrated model of supply chain assessment

A modular parametric approach can introduce a flexible and precise way to assess the relative contributions of scalable supply chains within a product chain.

Such an approach structures the available data (e.g., information on energy and material input; quantitative emissions into the water, soil, and atmosphere; transport data from suppliers to focal companies) in terms of input and output impacts for each product chain node.

Further simulation of supply chain trade-offs, which also account for other quantitative indicators, assign performance indicators to each supplier. Other reference indicators for such assessments are the delivery time, the quality of the product, the flexibility, the inventory strategies, and the environmental profile.

As reported in the gray boxes in Figure 11.2, firstly, product chain modularization provides the set of quantitative data; then, dynamic simulation integrates this information and provides quantitative values for unavailable data. With such an approach, the simulation can perform assessments for several products and supply chain configurations. A final analysis of customized production models allows one to assess the sustainability due to different manufacturing scenarios within the make-to-order paradigm.

11.3.1 Modular LCA approach for supply chain modeling

The modularization of the impact assessment starts from a comparative LCA. As a first step, LCA execution is compliant with the LCA guidelines (DIN EN ISO 14040:2006/14044:2006). LCA consists of four phases: (1) definition of goal and scope, (2) inventory analysis, (3) life cycle impact assessment (LCIA), and (4) interpretation.

The modular life cycle approach includes the definition of the examined system, functional units, system boundaries, allocation procedures, data quality requirements, and any other assumptions.

- *Goal and scope*: As opposed to traditional LCA scopes, which depend on a specific product and the intended use of the study, the modular approach aims to identify single information modules for each recurrent *macro flow* within the product chain. Macro flows are aggregated flows (i.e., specific products or services) commonly exchanged within the supply chain. According to the extent of optimization, both cradle-to-gate and cradle-to-grave perspectives can be adopted.
- *The LCI phase* is an inventory of input/output data with regard to the examined system involving the collection, calculation, and allocation of the necessary data. The modular life cycle requires tracking product chain data according to recurring flows for a wide range of possible products.
- *The LCIA phase* provides additional information to help assess a product system's LCI results in order to understand their environmental significance. The approach focuses on the environmental impact significance and the relative contribution of each flow. The results represent each individual impact from a comparative perspective.

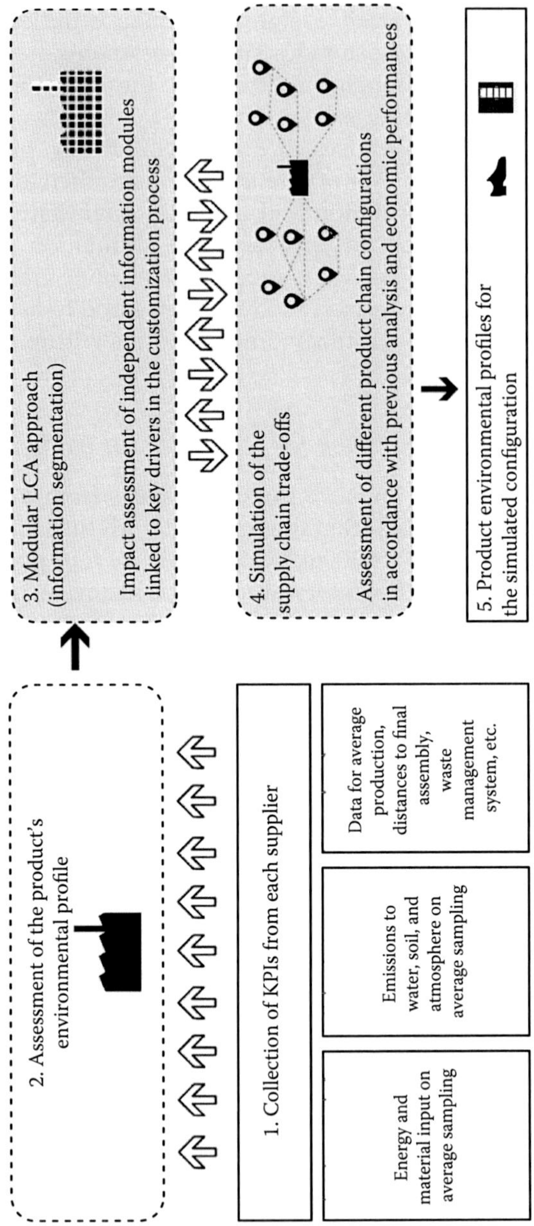

Figure 11.2 Integrated model for supply chain evaluation.

- *Life cycle interpretation* discusses the results of the LCI or the LCIA as a basis for conclusions, recommendations, and decision-making in accordance with the goal and scope definition. This phase compares single results in addition to other weighting factors such as times and cost drivers. The supply chain assessment model integrates the LCIA phase in order to identify the best supply chain configuration.

Equation 11.1 formalizes the modular approach in quantitative terms. Such quantification takes into account a series of modularizations for the traditional impact assessment.

- *Modularization of single life cycle phases*: The impact assessment of the product life cycle is the sum of the environmental profiles of different phases (i.e., manufacturing phases or supplier operations). In particular, the indexing of such modules identifies the supply level from one player to another (i.e., from $z-1$ to z). The LCIA of the customized product is the sum of incremental contributions from different supply chain stages.
- *Modularization of the customized product*: A certain number of physical components form the customized product with reference to a specific supply chain configuration (i.e., a material type from a specific supplier). The environmental impact assessment of the customized product is the sum of environmental profiles for each different component.
- *Explication of the key manufacturing drivers*: With reference to the specific customized product, the approach identifies and clusters relevant drivers with a significant variance and potential environmental impact. Common drivers due to product customization are the material composition of a product, the product weight, the product chain transport, the warehouse stock for each supplier, and the material waste of a single operation. The environmental impact assessment of the customized product is the sum of environmental profiles depending on such key drivers.

From a supply chain perspective, each node of the network represents a single company, while an input–output model defines the flow inventory of a single company (Hart, 1997).

In more detail, firstly, the life cycle analyst assesses the impact of background flows (auxiliary flows and processing materials) and foreground flows (final products and emissions to nature) for a certain company. Such an approach requires tracking and collecting the flows crossing the physical factory boundaries with reference to the final product. Common flows are input energy vectors (e.g., the electricity and natural gas used for the operation of the production plant), input primary resources, (e.g., the water supply), output emissions to air and water (e.g., volatile organic compounds

(VOCs), persistant toxic substances (PTS), and wastewater), and output solid waste with related treatment (e.g., wastes from packaging and finishing activities, paints, and coatings). After the identification of such recurrent flows, the life cycle analyst calculates the respective environmental impact for a reference unit in compliancy with the LCA general rules. The results constitute a set of independent information modules for a company.

Subsequently, in a further calculation, the life cycle analyst gathers together and calibrates the information modules according to the overall mass and energy balance related to the product chain activities. While the impact of elementary flows (e.g., basic chemicals and energy vectors) requires standard values from international databases, specific materials and components require specific LCA study or, alternatively, the adoption of the same approach as the foregoing supplier.

The final evaluation of the impact assessment is inferred from a combination of the company inventory and life cycle studies for specific industry flows and elementary flows.

The following equations formalize the modular decomposition of the LCA approach in order to flexibly express the environmental profiles of a customizable item.

Equation 11.1 assesses the impact categories of the customized product as the sum of independent, previously calculated vectors. The final array expresses the cradle-to-gate assessment of a specific product from the raw material extraction up to the factory gate. The calculation method appears to be compliant with the supplier perspective. The same approach assesses different manufactured products within the same factory through a flexible supply chain and distribution methodology (Figure 11.3). Notations for the equations are shown in Table 11.1.

$$e_{pn} = EP_{\text{Company}}\left(p_n, z\right) + EP_{\text{supply}}\left(p_n, z-1\right) \tag{11.1}$$

$$EP_{\text{Company}}\left(p_n, z\right) = \frac{m(p_n)}{M \cdot N} \cdot \left[\left[\sum_{i=1}^{I}\left(u_i \cdot \sum_{k=1}^{K} q_k\left(p_n\right)\right)\right]_{\text{Input}} + \left[\sum_{j=1}^{J}\left(u_j \cdot \sum_{f=1}^{F} q_f\left(p_n\right)\right)\right]_{\text{Output}}\right] \tag{11.2}$$

$$E_{\text{supply}}\left(p_n, z-1\right) = \sum_{s=1}^{S} q_s \cdot e_{ps} + \sum_{t=1}^{T} u_t \cdot \left(\sum_{r=1}^{R} l^t \cdot m_r^t\left(p_n\right) \cdot d_r^t\left(p_n\right)\right) \tag{11.3}$$

$$e_{pn} = EP_{\text{tare}}\left(p_n\right) + \sum_{d=1}^{D} X_d\left(p_n, c\right) \cdot u_d \tag{11.4}$$

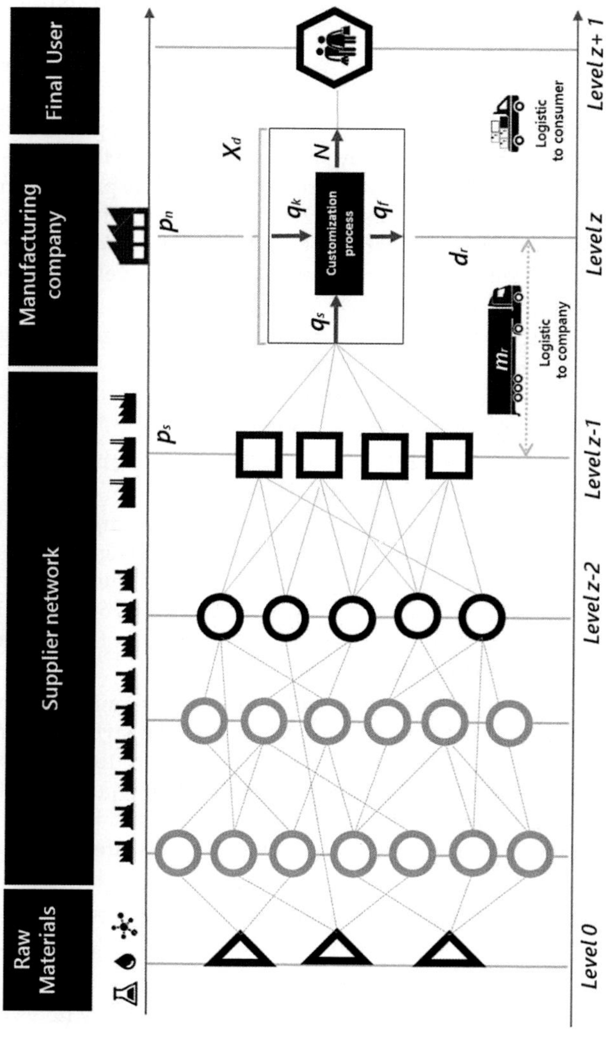

Figure 11.3 Modular representation of a product chain for a specific customized product.

Table 11.1 Notations for Equations 11.1 through 11.4

Variable	Explanation
e_{pn}	Environmental profile for the customized product (EPCP): the vectorial array of data representing environmental impact of the production of one customized product p_n by a specific company.
EP_{Company}	Environmental contribution to the environmental profile of the product p_n by the company internal processes. It represents the environmental impact due to consumption and emission of company activities.
EP_{Supply}	Environmental contribution to the environmental profile of the product p_n by the company supply. It represents the cradle-to-gate environmental impact of supplied items for the production of a reference quantity of customized products p_n in the reference period. The bill of materials of the customized product can help to list such items. The *supply* refers to a specific level (i.e., the direct supply to the company).
$m(p_n)$	Weight (or alternatively the value) of the customized product p_n in the reference period.
M	Total weight (or alternatively the value) of the total production of the company in the reference period.
N	Total number of the total customized products p_n manufactured by the company in the reference period.
u_i	Vectorial array representing the environmental impact for an incoming unitary flow of energy/mass. The inventoried mass and flows are *not* included in the final product p_n. This vectorial array refers to a homogenous flow type both in terms of physical features (e.g., the same energy input type) and in terms of product chain features (e.g., the same supplier).
u_j	Vectorial array representing the environmental impact of an outgoing unit of energy/mass flow type. The inventoried mass and flows are wastes changing with the kind of production (p_n). This vectorial array refers to a homogeneous flow type both in terms of physical features (e.g., the same waste type) and in terms of product chain features (e.g., the same dismissal procedure).
q_k	Inventoried quantity for a specific incoming flow type.
q_f	Inventoried quantity for a specific outgoing flow type.
I	Total number of incoming flow types.
J	Total number of output flow types.
K	Number of total supplies for the incoming auxiliary flows i by the examined company in the reference period.
F	Number of total disposals for the outgoing flows j from the examined company in the reference period.

(*Continued*)

Table 11.1 (Continued) Notations for Equations 11.1 through 11.4

Variable	Explanation
e_{ps}	Environmental profile for the supplied items (EPSI): a vectorial array of data representing the environmental impact of the supplied item p_s that composes the final customized product p_n.
S	Total number of supplied items for the production of the customized product p_n.
q_s	Quantity of supplied items p_s that are required for a single unit of the customized product p_n.
u_t	Vectorial array of the environmental impact of a specific transport type t. The vector is assessed for 1 ton*km and for a set of predetermined impact categories.
T	Total number of transport types.
l	Load factor for a single round trip.
m	Mass of the supplied items, to or from the company, that is transported by the transport type t.
d	Distance covered by the transport t in a round trip.
R	Total number of round trips between the suppliers and the company for the production of the product p_n.
D	Number of customization drivers changing during the customization process.
X	Value of the customization driver in a specific manufacturing scenario c (e.g., the quantity of a specific waste).
u_s	Vectorial array representing the environmental impact of a unit of a specific operational driver that changes during the customization process.
EP_{tare}	Total contribution to the environmental profile of the product p_n due to operational drivers that remain unchanged during the customization process.
p_s	Items and services provided from specific suppliers at a specific tier level.
p_n	Customized product manufactured in the examined company.

The approach can be applied both in the presence of previous cradle-to-gate LCA studies for specific product components and to further analyze the intermediate suppliers up to raw material level. Furthermore, the same approach can be applied from the consumer's perspective by a simple extension of the product chain to the level $z + 1$.

Finally, Equation 11.4 modularizes the same impact categories according to different operational drivers in order to introduce an explicit dependency of the LCA calculation from the production management choices.

11.3.2 Simulation of supply chain trade-offs

The output of the modular LCA is used for the second stage of the model. Discrete event simulation is used as a tool that enables one to evaluate alternative production network configurations and operating procedures in a convenient way when optimization models are not practical (Bernard et al., 2011). The model is developed to compare different scenarios with the initial configuration of the supply chain model. This part of the model gives companies the ability to create different configuration scenarios and to make what-if analyses to evaluate the trade-offs due to customization between different performance dimensions that are otherwise difficult to compare, such as delivery time and sustainability. For example, considering different customization policies, the need to shorten the delivery times to each customer can increase the number of deliveries, therefore increasing pollution. The model studies how to optimize the number of deliveries in the upstream supply chain without compromising delivery times to customers and without compromising sustainability. The model also evaluates the impact of applying different aggregations of orders to suppliers as a way to reduce their lead time and environmental impact.

The modeling of supply networks is often used as a way to check the balance of inventory, especially to compare standard production methods with just-in-time approaches. In the literature, three different approaches can be found: organizational, analytical, and simulation (Zhang et al., 2011). The first one relies on process modeling based on systems theory; however, the models developed with this approach are not dynamic and they do not take into account the system's behavior through time. The second one relies on mathematical formalization of the supply chains. These models, however, require approximations, usually restrictive, that can also be limited for considering time.

Simulation refers to a broad collection of methods and applications to mimic the behavior of real systems. Simulation models enable one to evaluate alternative system designs and operating procedures in a convenient way when the optimization of models is not practical due to the dimensions of the problem in terms of complexity. Moreover, simulation as support when testing alternatives on a real production system is usually too expensive and time consuming.

The model created for the specific case of comparing different customization strategies is based on the following starting points:

- The supply chain is based on a hierarchical relationship with the focal company: suppliers deliver to the company their materials and components on specific requests.
- Production orders are pulled by the customer orders; therefore, a make-to-order strategy is applied.

- It is assumed that there is one warehouse where all the materials and components are sent by the suppliers and are ready to be used according to the customer orders.
- Customer orders are received by the focal company and dispatched to suppliers with a fixed date policy and taking into consideration minimal safety stock.
- The customer orders are queued according to the request date from the customer with a first-in/first-out strategy.
- The performance of the suppliers is used to evaluate the overall performance of the supply network and is based on delivery time, quality (scraps), flexibility, and so on. For each of these indicators, the variance is also taken into consideration based on the real performance collected from the enterprise resource planning (ERP) of the focal company.
- Contractors are also part of the network structure—that is, companies working in parallel with the focal company when there is a capacity problem.
- The environmental profile is assigned to the three phases identified in the application of the modular LCA: suppliers, transports, and production at the focal company/contractors.

This model is modular and can be used and customized for different companies according to their specific data. Suppliers can be added according to the dimensions of the specific network and the performance adapted to the needs of the specific case.

11.3.2.1 Formulation of the model

The simulation allows one to verify the performance of different scenarios for each defined network configuration to analyze the effect of improving performance in the case of traditional or personalized products, and also considering the possibility of changing the number of suppliers and considering how much the overall performance will change when the performance of suppliers is improved.

Defining supplier i (where $i = 1,\ldots, n$) and order j (where $j = 1,\ldots, M$), the performance of each supplier is evaluated based on the following indicators:

- $T(i)$ is the delivery time of the supplier (i), evaluated as the average time to deliver an order. This performance is particularly relevant in customization since it is necessary to provide customized products in a short time, assuring flexibility.
- $Q(i)$ is the quality of the supplier (i), evaluated as the average percentage of defective pieces in each delivered order. This performance is particularly relevant in the customization context because defective

pieces are hardly tolerated by consumers willing to pay an even higher premium price for customized products; defective products create delays in delivery due to the required rework.

For what concerns the production orders that the focal company assigns to suppliers, their demand occurrence follows a normal distribution $N(\mu,\sigma)$, where μ is the mean of demand and σ is the standard deviation.

The simulation is replicated to create different supply chain configurations. Then, each configuration is evaluated based on the following supply chain performance indicators:

- *Order lead time* (OLT) is the time from receipt of the order from the customer (i.e., the focal company's retailer) that starts the supply chain production process to the delivery of the product to the customer (that is, the end of the supply chain process).
- *Inventory volume* (IV) is the volume of the inventories of components that are transferred from suppliers and used at the product factory.

The creation of comparative supply chain configuration scenarios (i.e., scenarios 1, 2, etc.) is based on the variation of the suppliers' performance starting from scenario 0. In the simulation model, the production costs are not considered because it is assumed that they are not a discriminant in the choice of customization since it is demonstrated that customers are willing to pay a premium price for customized products (Alptekinoglu and Corbett, 2008). Table 11.2 shows the to-be supply network configuration scenarios created with the simulation.

11.3.3 *Manufacturing scenarios*

The definition of different manufacturing scenarios allows direct assessment of the environmental implications of customization policies. This means identifying recurring customization in industry practices, the degree of variability of the product, and the degree of variability in the related supply chain.

The customization scenarios aim to fix the driver variance for a certain product batch in the presence of a progressive increase in the product variance toward the final consumer. Table 11.2 reports the general assumptions made in mass production and mass customization.

11.3.3.1 *Customization drivers*

Customization strategies can vary according to the combined variations of technical, market, and organizational drivers. The following list reports the relevant drivers according to previous literature studies and to an analysis of several companies dealing with customization.

Table 11.2 Manufacturing scenarios

Manufacturing scenarios	Description	Design changes	Supply chain changes
Mass production (current situation)	Supply matches a certain quantity with a minimal flexibility within a year. The modeling of material supply requires an average load per travel.	The bill of materials is fixed and design changes are limited to standard sizes and color changes.	Suppliers are located in various countries (according to the current location of the real case) according to cost and quality parameters.
Mass customization	Production follows the style preferences of the customer within a certain degree of freedom in the choice. This higher grade of selection includes a variability in design features, ergonomic features, material features, and aesthetical features.	The bill of materials can change in terms of component size and component type. The number of materials for the components and the number of suppliers increase by allowing the customer to change the material type within certain components for aesthetical, technical, or ergonomic reasons.	Accessories and components can be supplied in a local network (average distance from focal company is limited to a certain mileage), while suppliers of raw materials are kept unchanged. Possible variations of stored stock, extra consumption, and waste and transport depend on the consumer order sequence.

11.3.3.1.1 Operational drivers

Number of models within the same production batch: Starting from a specific type of product (e.g., shoes), the number of models available can vary per thousand of shoes produced. The change involves a limited improvement of ergonomic features and variations in the type of material for product components (in the case of shoes, it can be an increase in the types of leather for the upper and variations to the outer sole). The more extensively the customization is applied, the more the bill of materials changes in terms of components typology.

Processing materials and scrap rate: The increase in the variability of the final product can affect the efficiency of the traditional manufacturing process. In particular, the requirement of material per pair of shoes should include the gross material requirement. As an example, in comparison with mass production, customization can increase waste production with the concurrent manufacturing of different shapes for the upper within the same production batch.

Defectiveness rate: Product defects depend heavily on technological and managerial processes that exist within a single company. Although it is difficult to quantify the change in defectiveness levels with the product variability, it is possible to assume that such defectiveness contributes to the increased complexity of manufacturing options. Defective products can affect environmental impact due the additional resource consumption for a single product and with the increased waste contribution.

Transportation: The increase in the number of deliveries for a manufactured product seems to depend on supply chain management and the size of the production batches. In general, an increase in material types from different suppliers can imply a decrease in transport efficiency and in the load optimization. Such an effect can be registered both at the factory gate (more limited supplies) and in the output to consumer distribution (smaller lots at the points of sale).

Auxiliary material consumption: The consumption of materials and auxiliary resources (consumables not integrated into the final product) in general has limited dependence on the variability of the product. In fact, the consumption of auxiliary materials depends on an increase in the variety of the product only within a limited amount. Instead, the technology for the production process significantly infers the consumption and emissions for each type of model. However, the growing complexity of the production processes may entail a limited increase in these consumptions.

11.3.3.1.2 Economical drivers

Unsold items: Unsold items depend on the failure to predict the market demand. Despite the economic and environmental damage related to overproduction, the price elasticity of the demand for goods could reset the stocks of unsold items. In the case of customization, it is possible to assume that an increase in choice for consumers can better satisfy demand and reduce the unsold items.

Average product life cycle: Some economic studies (Brodie et al., 2013) suggest that increased demand satisfaction has a limiting effect on the replacement of an asset. There is a lack of empirical links between the increasing customization of a product and the reduction of its replacement. However, it is possible to assume (within the further assumption that the satisfaction remains the same during the product use) that customized products fit better with customer needs and may increase their time of use, reducing new consumption in a certain period.

Order size from selling points: Increasing market segmentation and increasing customization of products may increase the frequency of supplies to retailers, shops, and multistores. There is no reason to keep high stocks of customized items, and this can in fact be risky due to fluctuations in demand.

Time to service: In a scenario of stable technologies, a lack of optimization within the product chain is highly dependent on the required time to service and the demand trend. Segmented markets with high variability may in fact require rapid production organization, with implications for the demand of related resources and environmental emissions.

11.3.3.1.3 Organizational drivers

Make-to-order supply chain: A chain of suppliers that is organized according to the lean *make-to-order* paradigm with a reduced stock at the final assembler and a frequent supply depending on the customized product demand. This chain type requires efficient organization and a restrained time to market. Transportation can remain frequent and nonoptimal even if the assembler and suppliers are synchronized.

Factory flexibility: A flexible factory is able to meet a variable demand for customized products and a proper time to market. In order to perform such operations, the factory includes many production departments and an adequate internal materials stock.

11.4 Sustainability assessment for a customization case in a fashion company

11.4.1 Application of modular LCA

The application of a modular LCA to a footwear case enabled a comparative assessment of environmental burdens due to customization policies in a fashion company. The LCA involved a cradle-to-gate perspective on the footwear company, and the analysis took into account the product supply chain from raw materials acquisition up to product manufacturing and industrial waste disposal. The product's use and its dismissal were not included in the model. The modeling of the factory waste also included the waste treatment processes after the initial deposit. The analysis did not take into account the waste flows sent for economic recovery (e.g., material recycling, energy recovery, and composting). In this case, system boundaries were limited up to the facility gate where the recycling or recovery processes take place (i.e., transportation to the facility was included).

1. In the first stage, a classical LCA assessed the common recurrent flows for an Italian footwear company. The combination of such recurrent flows provided the total environmental profile for the factory in the reference period. Such impact results from the combination of the industry flows (e.g., the average energy used for each shoe pair) and the processing materials (e.g., the specific content of material per footwear type).
2. The use of data from international databases (e.g., Ecoinvent and Gabi) supported the LCA model, particularly for elementary flows. The formalization of environmental impact through impact categories is compliant with the CML 2011 standard and environmental product declaration (EPD) system: specified in General programme instructions for the International EPD System, 2013. The impact categories used to assess inventory flows were global warming potential (GWP), acidification potential (AP), eutrophication potential (EP), ozone depletion potential (ODP), and photochemical ozone creation potential (POCP).
3. In the second stage, the modular LCA assessed the environmental impact variance due to the customization of a production batch under specific conditions. A number of company drivers address the variance assessment. In terms of technology options, we assumed that the production of customized footwear required the same resources as the current technologies.

11.4.2 *Life cycle inventory for the case study*

The methodology of data collection, compliant with the modular approach, allowed the acquisition of data sheets and inventory data from the examined company. At the factory level, the data accounts for mass and energy recurrent flows for an average yearly production of 477,569 footwear pairs.

Specific energy supply configurations referred to the energy mix of the utility serving the company (e.g., kWh supplied by a specific utility). In addition, the modeling of waste treatments complies with the European

Waste Catalogue (EWC) for industrial waste (e.g., recycled paint and varnish containing organic solvents).

The data inventory for different footwear models allowed the identification of the general impact for average footwear (Figure 11.4). Supply scenarios integrated the number of deliveries within a year, the distance between suppliers and factory, the means of transport, and the load capacity for each supply type (see Figure 11.5). The stages from resource extraction up to the creation of process materials involve suppliers from large distances (Level 1 in Figure 11.5). For example, the production of leather requires breeding outside Europe, transport to European tanneries, and then the manufacturing of the materials. The stages from the acquisition of process materials to shoe manufacturing involve manufacturing at local levels, so the producers of components and the footwear manufacturing company are placed in a local district over 100 km (see Level 2 in Figure 11.5). We lastly assume that the final footwear consumer stays in the local area (less than 200 km) (see Level 3 in Figure 11.5).

Different drivers are considered according to Equation 11.4 in order to better define the scenarios to be analyzed (Figure 11.6).

A description of the product's physical features, the supply chain configuration, and the manufacturing features are reported in Table 11.3. The selection of these drivers defines a basic scenario in which each operational driver has a base value.

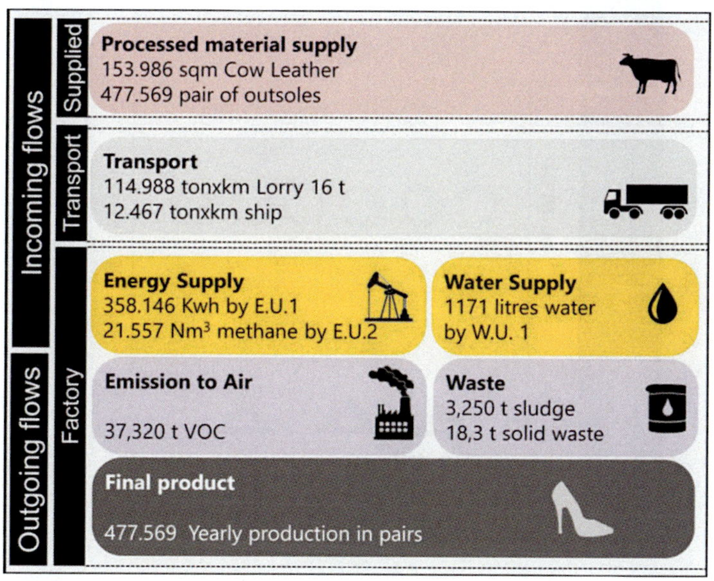

Figure 11.4 Inventory data for the definition of in- and outgoing flows.

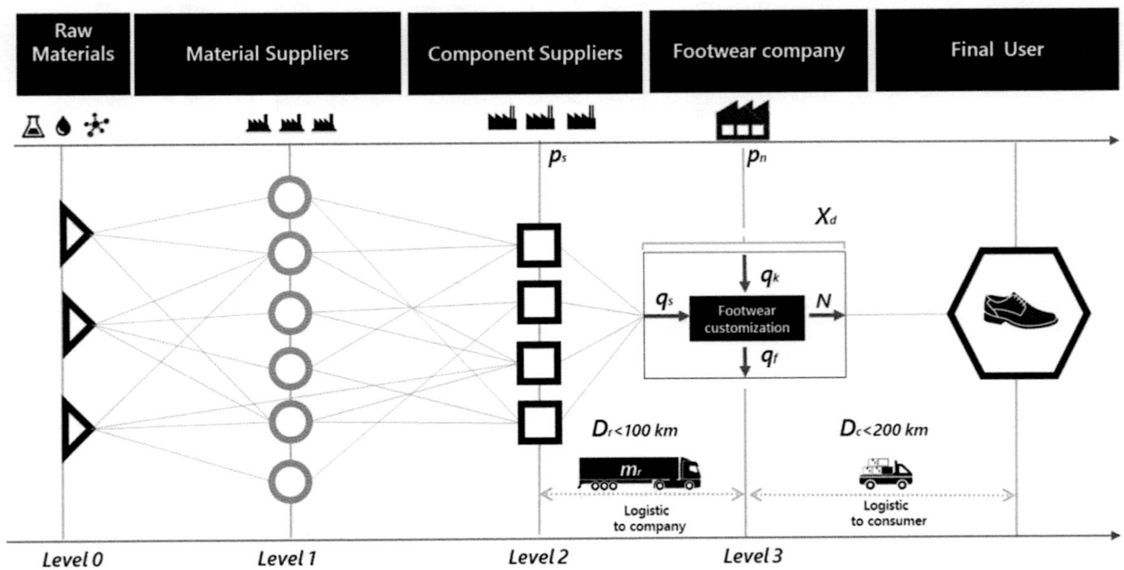

Figure 11.5 Model application to the study case.

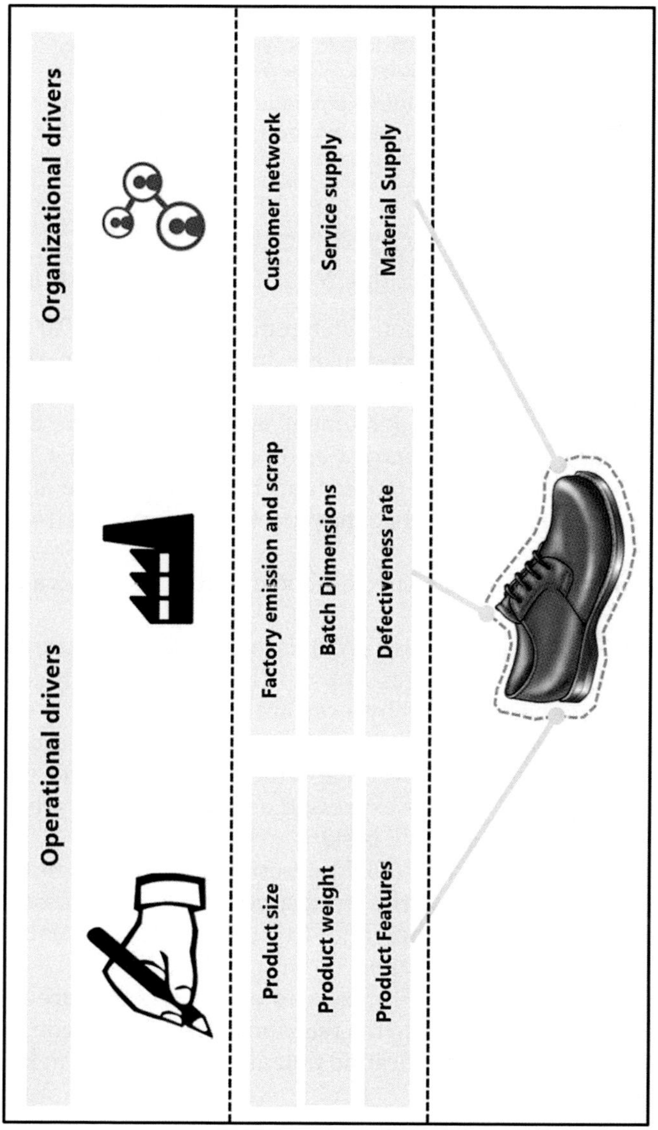

Figure 11.6 Operational drivers for the product customization.

Table 11.3 Product features

Product features	Sole	The sole is composed of ethylene-vinyl acetate and has a weight of 520 g
	Upper	The upper is composed of leather and has a total area of 1.5 m^2/pair
	Material composition	Leather upper, polyurethane sole, other materials—polyester, nylon 6, elastane, spandex, conventional cotton, rayon viscose, ethylene-vinyl acetate
	Pair weight	920 g

11.4.3 *Environmental impact due to customization activities*

In the assumptions, the production batch remains constant (10,000 pairs), while changes in the product design are introduced within the same batch. It is assumed that each driver varies according to a range that has been defined for this work, in agreement with the literature data and empirical evidence. Each driver varied as reported in Table 11.4 and multiple impacts were analyzed to understand how the positive and negative influence of different drivers can impact the overall environmental performance.

In particular, the following environmental indicators are evaluated:

- Global warming potential (GWP) is a relative measure of how much heat a greenhouse gas traps in the atmosphere. It compares the amount of heat trapped by a certain mass of the gas in question with the amount of heat trapped by a similar mass of carbon dioxide. A GWP is calculated over a specific time interval: for example, 100 years (GWP_{100}). GWP is expressed as a kilogram of carbon dioxide equivalents (whose GWP is standardized to 1).
- Ozone depletion potential (ODP) describes the decline in the total amount of ozone in the earth's stratosphere. ODP is expressed as the sum of ozone-depleting potential in kilogram CFC-11 equivalents (e.g., over a period of 20 years).
- Acidification potential (AP) measures acid gases that are released into the air or resulting from the reaction of the nonacid components of the emissions. The acidification potential is expressed in kilogram sulfur dioxide equivalents.
- Photochemical ozone creation potential (POCP) measures the emissions of gases that contribute to the creation of ground-level ozone. The POCP is expressed in kilogram ethene equivalents.
- Eutrophication potential (EP) measures the ecosystem's response to the addition of artificial or natural nutrients, mainly phosphates,

Table 11.4 Description of customization drivers and the related variance range

Driver		Description	Initial value	Maximum value
E_1	Scrap rate	The scrap rate can vary according to the efficiency of the cutting process for the upper. Such efficiency can be reduced by the alternation of different models within the same factory. A single supply could be oversized with respect to the batch size. The base value is based on "Waste generated in the leather products industry," (Schmél, 2000).	20% with respect to the bill of material values	60% with respect to the bill of material values
E_2	Transports	Transport depends on the number of deliveries to the factory, the distance between producer and suppliers, and freight load factors. According to the case study, the components supplier remains in an area with a radius of 50 km. Then the supply numbers increase. In the base scenario, an average truck type performs the supply in a single trip for the whole batch (9200 kg).	50 km	1000 km
E_3	Defectiveness rate	In the case study, the defectiveness rate can increase with the manufacturing complexity. Each defective pair requires a new production to maintain the batch dimensions. In the base scenario (mass production), the defectiveness amounts to 10 pairs per 10,000 shoe pairs.	0.1% with respect to the batch size	10% with respect to the batch size
E_4	Life cycle extension	This driver introduces a positive effect. Life cycle extension reduces product replacement that creates a minor provision of further resources from the eco-sphere in a certain period time. According to assumptions, the shoe pair can last more than 2 years while its effective average use amounts to 2 years.	2 years' duration (expected use phase)	3 years' duration
E_5	Material variance	The overall batch size is 10,000 pairs. The uppers of the shoe are made of different materials within the same production batch and for the same model. Every time a new material is added, the batch is split into smaller sub-batches. For example, two material types means that 5000 pairs are produced with one material and another 5000 pairs are produced with the other material type; three material types means 3.333 pairs per material type, and so on.	One material type for the upper (leather)	22 material types for the upper (5 leather types and 17 synthetic types)

through detergents, fertilizers, or sewage, to an aquatic system. The emission of substances to water contributing to oxygen depletion is then expressed as kilogram phosphate equivalents.

Appendices 11.1 through 11.5 report, respectively, the environmental impact due to customization activities for the five environmental indicators, GWP_{100}, AP, EP, ODP, and POCP.

It is important to emphasize that the results were calculated under the assumption that the technology framework remains the same, assuming that an increase in the level of customization is linked to an increase of the product model variability.

In the charts in the annexes, the variations of all the operational drivers mentioned in Table 11.4 (life cycle extension, scraps, defectiveness, transport, material variance) are normalized to a scale 1–100 to make them comparable. The starting situation is represented by a blue bar and all the bars above it represent the cases when the variations of the five drivers are such that they cause an increase in the environmental impact, while the bars below it represent a decrease in the environmental impact.

The results for the specific case study suggest the following conclusions:

- The customization process can have both positive and negative environmental impacts, and when it is linked to an increase in the possibility of using new materials, these may in fact include more eco-efficient than traditional materials and thus reduce the environmental impact.
- The most significant drivers to control the environmental impact are the choice of material type, the rate of defective parts, and the scrap rate.
- The positive effect on the environmental impact brought about by the life cycle extension of the use phase of the product, avoiding the use of new resources, balances increases in other drivers. Similarly, a more informed consumer choice on the test material could affect the impact on the final product.
- Impact categories are affected differently by the product variance; in the analyzed case, AP and ODP indicators doubled their impact according to the change of the operational parameters.
- The allocation rules can also significantly affect the background impacts. Standardization plays a role in determining the best supplier options for foreground sectors. A clear alignment between allocation rules and system boundary selections with respect to background suppliers (material producers) seems necessary in order to reduce the potentially high variability in LCA results.

11.4.4 Results of the simulation of customization scenarios

The simulation was based on the data collected from the ERP of the company and on the results of the modular LCA. Preliminary analysis of the data extracted from the ERP shows that the suppliers of the shoe company are asked to produce both large and small orders according to the needs of the company, with a wide range of order dimensions both in terms of the number of rows, the number of pieces per row, and the number of different items. According to the order dimensions, suppliers have different performances in terms of delivery time, product quality, and so on (see an example in Table 11.4). Before applying the simulation model, a Pareto analysis allowed suppliers to be categorized to identify the most strategic ones in terms of the total delivered amounts. In the case study, it emerges that some supplier performances, such as average delivery time, are linked to the order dimensions, while others are independent from them, such as average scraps. These performance indicators are taken into consideration in the simulation model and are used to create the scenarios for large and small orders.

According to the defined model, the customization strategies have been applied to choose the most suitable suppliers for each scenario, and a commercial simulator (Simio®) was used to compare different scenarios based on suppliers' performances. The initial scenario was based on data collected from the footwear company, and it represents a simplified model of its network where most of the suppliers are considered. The model is based on the following assumptions:

- A contractor works in parallel with the shoe producer to manufacture the orders that can't be assembled by the shoe producer due to capacity limits.
- Some product models can be produced only by the shoe producer, others only by the contractor, and others by both of them.
- In cases where a product can be processed both by the shoe producer and by the contractor, it is sent to the one with the shortest queue.
- The shoe producer manages the materials necessary for the contractor and forwards them when necessary for production.
- The warehouse and the distribution center are located at the shoe producer's site.
- The working time of the contractor includes extra time both for the delivery of materials to the contractor and for shipping the final products to the distribution center.

The advantage of producing at the contractor is given by the fact that there is the possibility to shorten the queue of the company.

The application of the modular LCA in the previous section showed that out of the three identified macrocategories (supplying process, transportation, production), the supplying process has a large impact on the overall sustainability of the network, and for this reason the scenarios are built mainly to evaluate how their performance can impact on the sustainability—in particular, considering the most important environmental indicator, the GWP.

As was described in the previous section, the scenarios defined in the LCA are used to link the level of customization (in terms of the number of product variations) with drivers such as transportation, scraps, defectiveness, and so on. The model is based on the same type of raw material (leather) being provided by the same suppliers or by similar suppliers to evaluate the environmental impact of their operative performance, besides the environmental impact of new materials. Therefore, starting from the standard production scenario, the other scenarios are analyzed according to possible changes in the operative performances of the suppliers, given the materials they can provide. A set of different what-if scenarios based on variations in the suppliers' performances has been defined in order to evaluate how changes in supplier performance can impact overall supply chain performance. In particular, it has been analyzed how improvements in their delivery time (from 10% to 35% of suppliers' lead time) and in product quality (from 10% to 35% in scraps) can affect the overall performance of the supply chain. Based on the data collected from the company and the established model, the performance of the supply network is dynamically evaluated, considering the value of the initial inventory, the average inventory during the analyzed period, and the average and maximum lead times to fulfill customers' orders.

Preliminary results show that variations in the suppliers' lead times have a different impact according to the applied level of customization (i.e., the number of product variants). Figure 11.7a shows how an improvement in supplier lead time performance can bring about an improvement in the customer order time, which is more than the improvement caused by supplier quality in the product defectiveness represented in Figure 11.7b.

In fact, as for the impact of changes in products delivered by suppliers in terms of quality (less scraps), with the data of the specific company it turns out that a reduction in scraps gives a reduction in final product defectiveness. The level of scraps represents a limited share of production (5%), and for this reason the impact is more limited than in the case of lead time changes. An improvement in the scraps level of product components means less reworking and less mistakes that go from the suppliers to the final customer, and means less defectiveness during production. Generally speaking, improvements in the suppliers' performances bring different degrees of improvement to the overall supply chain performance, and many variables need to be taken into consideration. In this

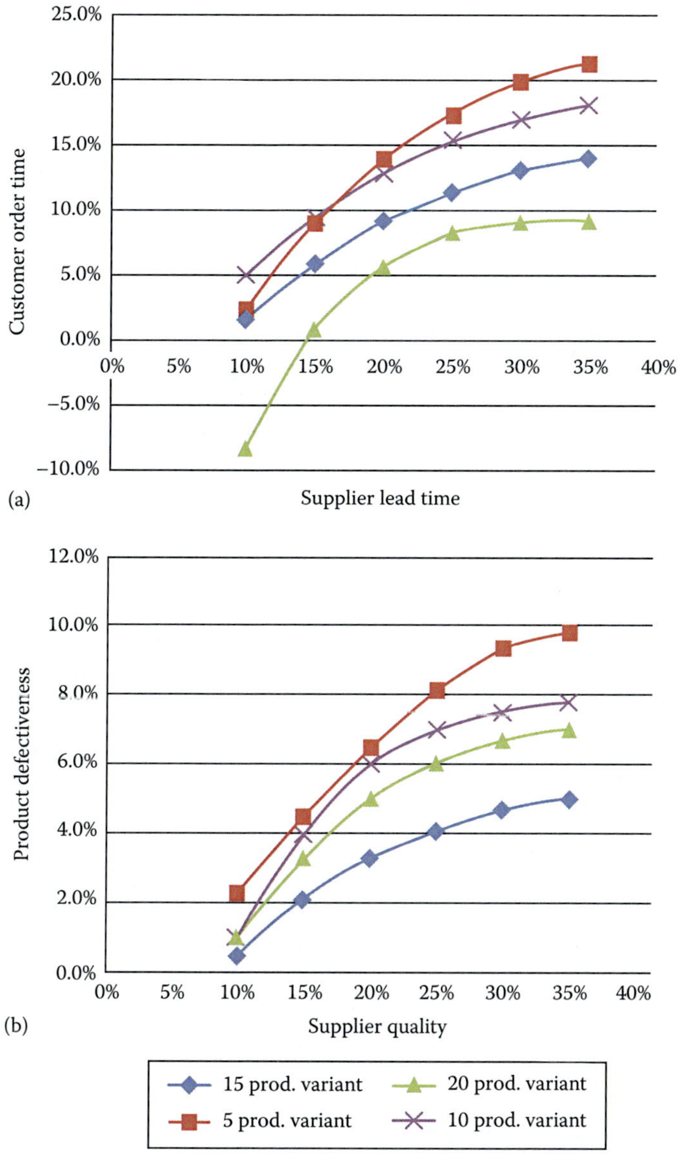

Figure 11.7 (a) Impact of improvement in supplier lead time on improvement in the customer order time. (b) Impact of improvement in supplier quality on decreasing product defectiveness.

study, some of them have been considered and analyzed, but further studies will be necessary to complete the flow.

11.5 Conclusions and recommendations

The complexity of evaluating the environmental impact of supply chain modeling seems to require novel methodologies to properly identify the key decisional areas. In this chapter, a new approach has been presented based on the integration of LCA data with discrete simulation and has been tested in a specific case by collecting data from a footwear company and considering different customization strategies.

The functional unit for manufacturing a product is commonly based on a single product. By adopting a factory perspective, it seems necessary since the LCI to shift the focus onto production batches rather than a single product. Such a shift could in fact include new inventory categories that represent more precisely the real hidden flows of customized production. Examples are the modeling of the distribution platform or the use pattern for a certain product. When the inventory is based on a batch, such an evaluation could include new variables more in line with mass and energy balance at the supply chain level.

The chapter also analyzes variances due to different product chain configurations in the environmental impact, based on the simulation of multiple scenarios considering different degrees of variability of the operational drivers. In the preliminary results, it is highlighted that specific decisional areas under the control of product managers are also key drivers in environmental impact creation. Further studies in other sectors could better contextualize the environmental implications. In particular, aspects such as economies of scale, warehouse management, and the use of alternative technologies could significantly affect this analysis.

The outcome of the model suggests that the proper implementation of customization practices could result in an environmental benefit. In general, it is possible to identify four subsequent scenarios for the implementation of sustainable customization practices (Figure 11.8).

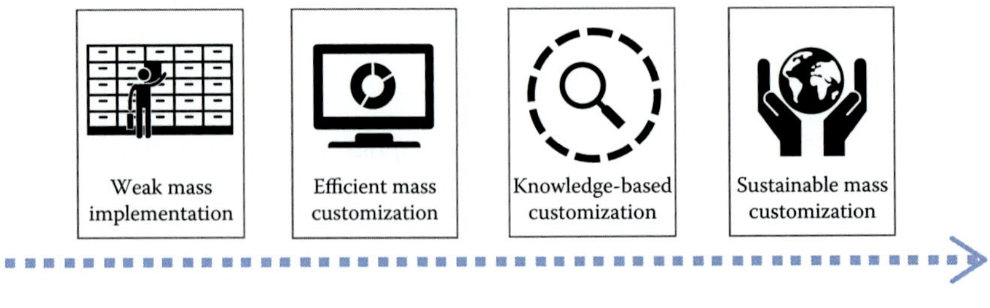

Figure 11.8 Scenarios for implementation of customization within the product chain.

- In the first scenario, *weak implementation*, customization focuses on limited aspects of the product such as design or some functional parameters without any framework to support consumption reduction or emissions. In this case, it becomes likely that the customized product will have a higher environmental impact.
- In the second scenario, *efficient customization*, dedicated tools can be implemented in a single factory perspective to minimize customization costs and consumption. In particular, emphasis is given to methods of effects quantification and data management from the manufacturer.
- In the third scenario, *knowledge-based customization*, personalization pushes onto multiple aspects concerning the use phase of the product and background phases so that the data concerning the whole product life cycle can be analyzed by the producer. This type of implementation makes clear the effects induced on the product chain and acts proactively to reduce these effects from a single-player perspective.
- In the fourth and final scenario, *sustainable customization*, data regarding the sustainability of the product is exchanged within the product chain with a standard protocol. The diffusion process involves the whole chain, starting from raw material producers up to the final consumer. Furthermore, distributed methods and tools for the quantification of the social and environmental effects related to the choice of customization concurrently support the product chain players (e.g., consumer, producer, material developer) at each stage. The diffusion of this type of information introduces emergent properties and feedback within the system. Such a framework, jointly with the increased decisional power of the buyer, can directly link the product's environmental profile with customization preferences.

Modeling based on simulation was used because it offers a realistic observation of supply chain behavior and allows an analysis of supply chain dynamics. It provides an observation of the behavior of the network over time, to understand the organizational decision-making process, analyze the interdependencies between the actors of the chain, and analyze the consistency between the coordination modes and the decisional policies. Moreover, simulation can be coupled with an optimization approach, to validate the relevance and the consequences of its results.

Future developments in the model will be based on making available for companies reliable libraries of environmental impacts and on refining the simulation model to ease what-if analysis. Further analysis of the trade-offs between operative and sustainable performance is also necessary. From this perspective, the authors will further develop and customize the framework for other specific industrial case studies, with the definition of transversal methods and tools for sustainability performance analysis. The relationships between critical processes, improvement actions, and sustainability dimensions as well as suitable indicators will be deepened and updated.

References

Alptekinoglu A., Corbett C. J. 2008. Mass customization vs. mass production: Variety and price competition. *Manufacturing & Service Operations Management*, 10(2): 204–217.

Bernard A., Daaboul J., Laroche F., Da Cunha, C. 2011. Mass customisation as a competitive factor for sustainability, in enabling manufacturing competitiveness and economic sustainability. In ElMaraghy H. A. (Ed.) *Proceedings of the 4th International Conference on Changeable, Agile, Reconfigurable and Virtual Production (CARV 2011)*, pp. 18–25. University of Windsor, Ontario, Canada.

Boër C. R., Pedrazzoli P., Bettoni A., Sorlini M., 2013. Sustainability assessment model. In Boër C. R., Pedrazzoli P., Bettoni A., Sorlini (Eds.) *Mass Customization and Sustainability*, pp. 33–142, London: Springer.

Bojarski D., Laínez J. M., Espuna A., Puigjaner L. 2009. Incorporating environmental impacts and regulations in a holistic supply chains modeling: An LCA approach. *Computers and Chemical Engineering* 33: 1747–1759.

Brodie R. J., Ilic A., Juric B., Hollebeek L. 2013. Consumer engagement in a virtual brand community: An exploratory analysis. *Journal of Business Research* 66: 105–114.

Brondi C., Fornasiero R., Vale M., Vidali L., Brugnoli F. 2012. Modular framework for reliable LCA-based indicators supporting supplier selection within complex supply chains. *APMS* (1): 200–207.

Brondi C., Fragassi F., Pasetti T., Fornasiero R. 2014. Evaluating sustainability trade-offs along supply chain. *Engineering, Technology and Innovation (ICE), 2014 International ICE Conference on*, pp. 1–8 Bergamo, Italy, June 23–25.

Bruno T. D., Nielsen K., Taps S. B., Jorgensen K. A. 2013. Sustainability evaluation of mass customization. *IFIP Advances in Information and Communication Technology* 414: 175–182.

Christmann P. 2000. Effects of best practice of environmental management on cost advantage: The role of complementary assets. *Academy of Management Journal* 43(4): 663–680.

Clark R. A. 1994. The challenge of going green. *Harvard Business Review* 72(4): 37–38.

Dyllick T., Hockerts K. 2002. Beyond the business case for corporate sustainability. *Business Strategy and the Environment* 11(2): 130–141.

Florida R. 1996. Lean and green: The move to environmentally conscious manufacturing. *California Management Review* 39(1): 80.

Gallouj F., Weber K. M., Stare M., Rubalcaba L. 2015. The futures of the service economy in Europe: A foresight analysis. *Technological Forecasting and Social Change* 94: 80–96.

Govindan K., Azevedo S. G., Carvalho H., Cruz-Machado V. 2014. Impact of supply chain management practices on sustainability. *Journal of Cleaner Production* 85: 212–225.

Hagelaar G. J. L. F., van der Vorst J. G. A. J. 2002. Environmental supply chain management: Using life cycle assessment to structure supply chains. *International Food and Agribusiness: Management Review* 4: 399–412.

Handfield R., Walton S. V., Sroufe R. 2002. Applying environmental criteria to supplier assessment: A study in the application of the analytical hierarchy process. *European Journal of Operational Research* 141: 70–87.

Hart S. 1997. Beyond greening: Strategies for a sustainable world. *Harvard Business Review* 75: 66–76.

Hugo A., Pistikopoulos E. N. 2005. Environmentally conscious long-range planning and design of supply chain networks. *Journal of Cleaner Production* 13: 1428–1448.

Humphreys P., McIvor R., Chan F. 2003. Using case-based reasoning to evaluate supplier environmental management performance. *Expert Systems with Applications* 25: 141–153.

International EPD® System. 2013. General programme instructions for the International EPD System. Version 2.0, Published June 4th, 2013, Internal document of The International EPD® System, www.environdec.com. Dated June 4.

ISO/TS 14072. 2014. Environmental management: Life cycle assessment; Requirements and guidelines for organizational life cycle assessment. First edition published 2014-12-15, Accesible at https://www.iso.org/obp/ui/#iso:std:iso:ts:14072:ed-1:v1:en

Klassen R. D., Whybark C. D. 1999. The impact of environmental technologies on manufacturing performance. *Decision Sciences* 30(3): 599–615.

Klöpffer W. 2012. The critical review of life cycle assessment studies according to ISO 14040 and 14044: Rigin, purpose and practical performance. *The International Journal of Life Cycle Assessment*, 17(9): 1–7.

Kohtala, C. Addressing sustainability in research on distributed production: An integrated literature review. *Journal of Cleaner Production* 106(2015): 654–668.

Lee Y., Huang F. 2011. Recommender system architecture for adaptive green marketing. *Expert Systems with Applications* 38: 9696–9703.

Mourtzis D., Doukas M. 2014. Design and planning of manufacturing networks for mass customisation and personalisation: Challenges and outlook. *Procedia CIRP* 19: 1–13.

Mourtzis D., Doukas M., Psarommatis F. 2013. Design and operation of manufacturing networks for mass customisation. *CIRP Annals: Manufacturing Technology* 62: 467–470.

Mourtzis D., Doukas M., Psarommatis F. 2014. A toolbox for the design, planning and operation of manufacturing networks in a mass customisation environment. *Journal of Manufacturing Systems*. 36: 274–286.

Noci G. 1997. Designing green vendor rating systems for the assessment of a supplier's environmental performance. *European Journal of Purchasing and Supply Management* 2: 103–114.

Nwe E. S., Adhitya A., Halim I., Srinivasan R. 2010. Green supply chain design and operation by integrating LCA and dynamic simulation. In Pierucci S., Buzzi Ferraris, G. (Eds.) *Computer Aided Chemical Engineering* 28: 109–114.

Osorio J., Romero D., Betancur M., Molina A. 2014. Design for sustainable mass-customization: Design guidelines for sustainable mass-customized products. *Engineering, Technology and Innovation (ICE), 2014 International ICE Conference on*, pp. 1–9. IEEE.

Petersen T. D., Jørgensen K. A., Nielsen K., Taps S. B. 2011. Is mass customization sustainable? *MCPC 2011 World Conference on Mass Customization and Personalization*. pp. 162–168, University of Novi Sad.
Piplani R., Pujavan N., Ray S. 2007. Sustainable supply chain management. *International Journal of Production Economics* 111(2): 193–194.
Porter M. E., Van der Linde C. 1995. Toward a new conception of the environment competitiveness relationship. *Journal of Economic Perspectives* 9(3): 97–118.
Reinhardt F. L. 1999. Bringing the environment down to earth. *Harvard Business Review* 77(4): 149–157.
Romero D., Cavalieri S., Resta B. 2014. Green virtual enterprise broker: Enabling build-to-order supply chains for sustainable customer-driven small series production. *IFIP International Conference on Advances in Production Management Systems*, pp. 431-441. Berlin: Springer.
Rusinko C. 2007. Green manufacturing: An evaluation of environmentally sustainable manufacturing practices and their impact on competitive outcomes. *IEEE Transaction on Engineering Management* 54(3): 445–454.
Russo M. V., Fouts P. A. 1997. A resource-based perspective on corporate environmental performance and profitability. *Academy Management Journal* 40(3): 534–559.
Schmél, F. 2000. Wastes in the leather products industry. 14th Meeting of the UNIDO Leather Panel, Zlin, Czech Republic, 13–15 December 2000.
Seuring S., Müller M. 2008. From a literature review to a conceptual frame work for sustainable supply chain management. *Journal of Cleaner Production* 16: 1699–1710.
Su M., Chen C., Yang Z. 2015. Urban energy structure optimization at the sector scale: Considering environmental impact based on life cycle assessment. *Journal of Cleaner Production*, 112(2): 1464–1474.
Trappey A. J. C., Trappey C. V., Hsiao C., Ou J. J. R., Chang C. 2012. System dynamics modelling of product carbon footprint life cycles for collaborative green supply chains. *International Journal of Computer Integrated Manufacturing* 25(10): 934–945.
Ueda K., Takenaka T., Váncza J., Monostori L. 2009. Value creation and decision-making in sustainable society. *CIRP Annals* 58(2): 681–700.
UNEP (United Nations Environment Programme). 2015. Guidance on organizational life cycle assessment. UNEP. Available at http://www.lifecycleinitiative.org/wp-content/uploads/2015/04/o-lca_24.4.15-web.pdf.
Walley N., Whitehead B. 1994. It's not easy being green. *Harvard Business Review* 72(3): 46–52.
Yu V., Ting H., Wu Y. J. 2009. Assessing the greenness effort for European firms: A resource efficiency perspective. *Management Decision* 47(7): 1065–1079.
Zhang Y., Luximon A., Ma X., Guo X., Zhang M. 2011. Mass customization methodology for footwear design. *Digital Human Modeling* 6777: 367–375.
Zhu Q., Sarkis J., Lai K. L. 2008a. Confirmation of a measurement model for green supply chain management practices implementation. *International Journal of Production Economics* 111: 261–273.
Zhu Q., Sarkis J., Lai K. L. 2008b. Green supply chain management implications for closing the loop. *Transportation Research Part E* 44: 1–18.

Appendix 11.1 Global warming potential (GWP_{100})

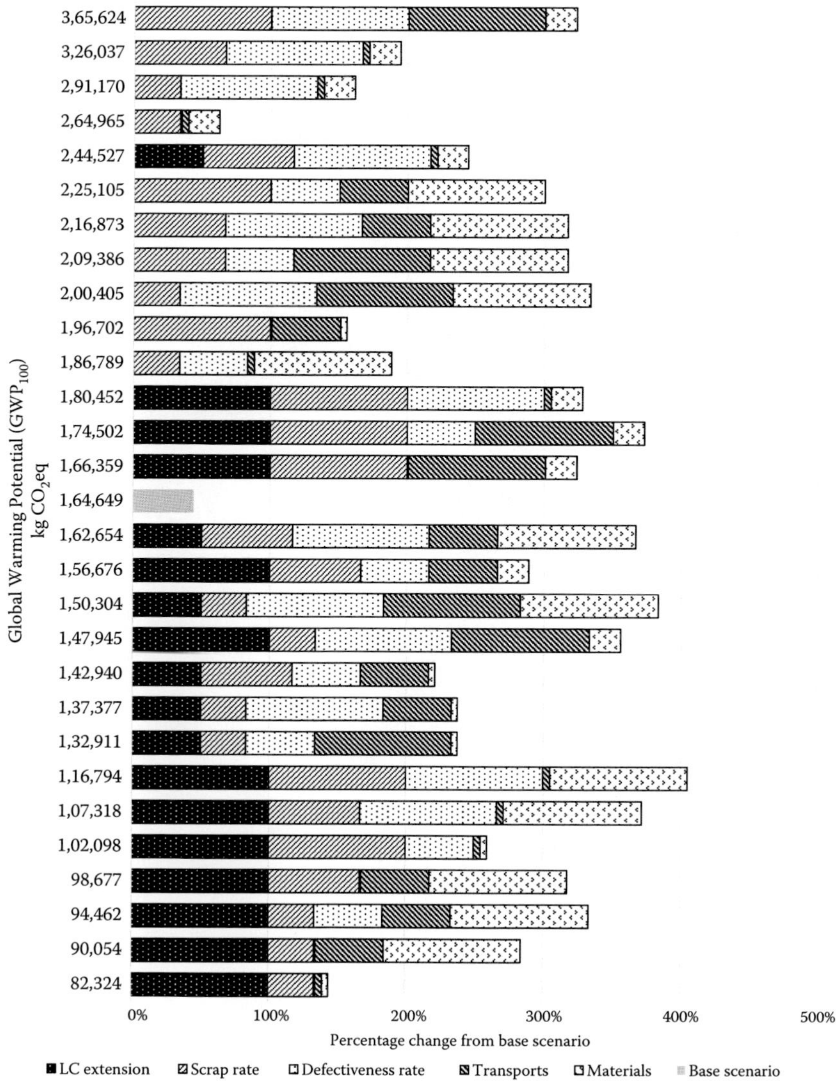

Appendix 11.2 Acidification potential (AP)

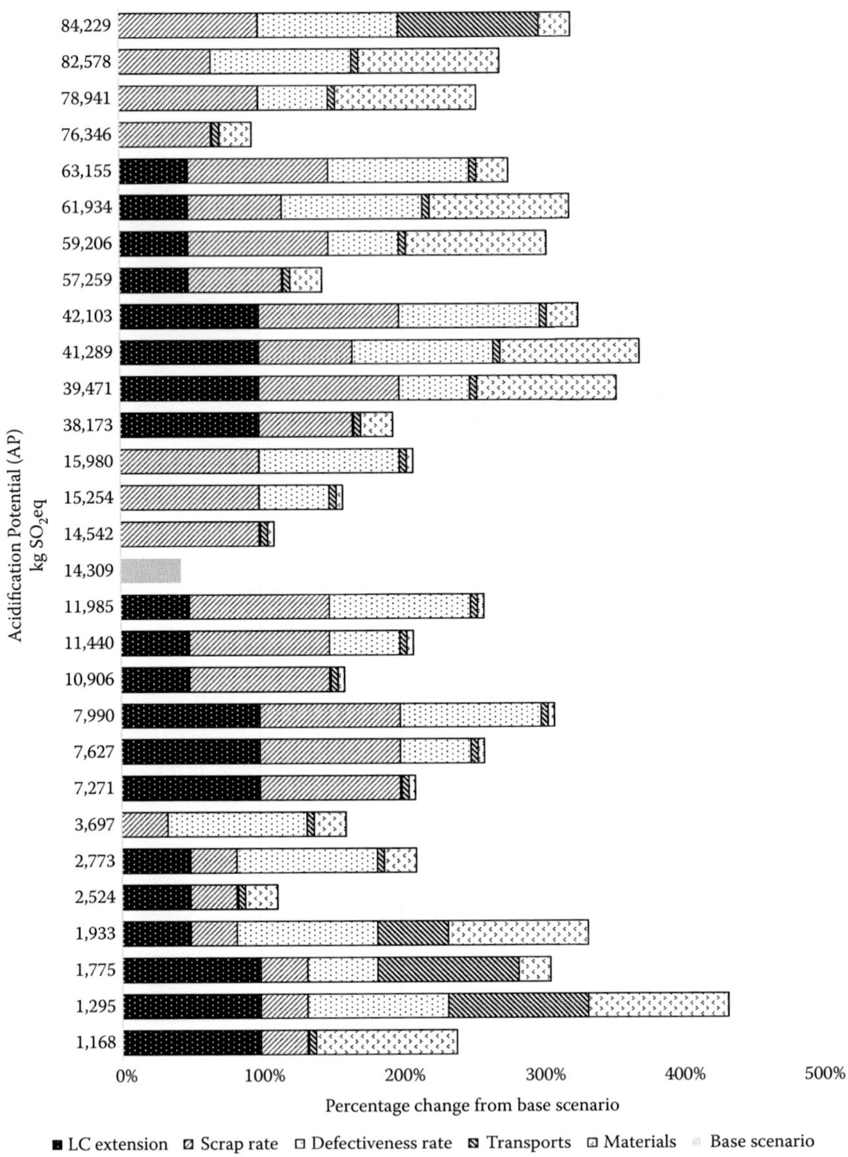

Appendix 11.3 Eutrophication potential (EP)

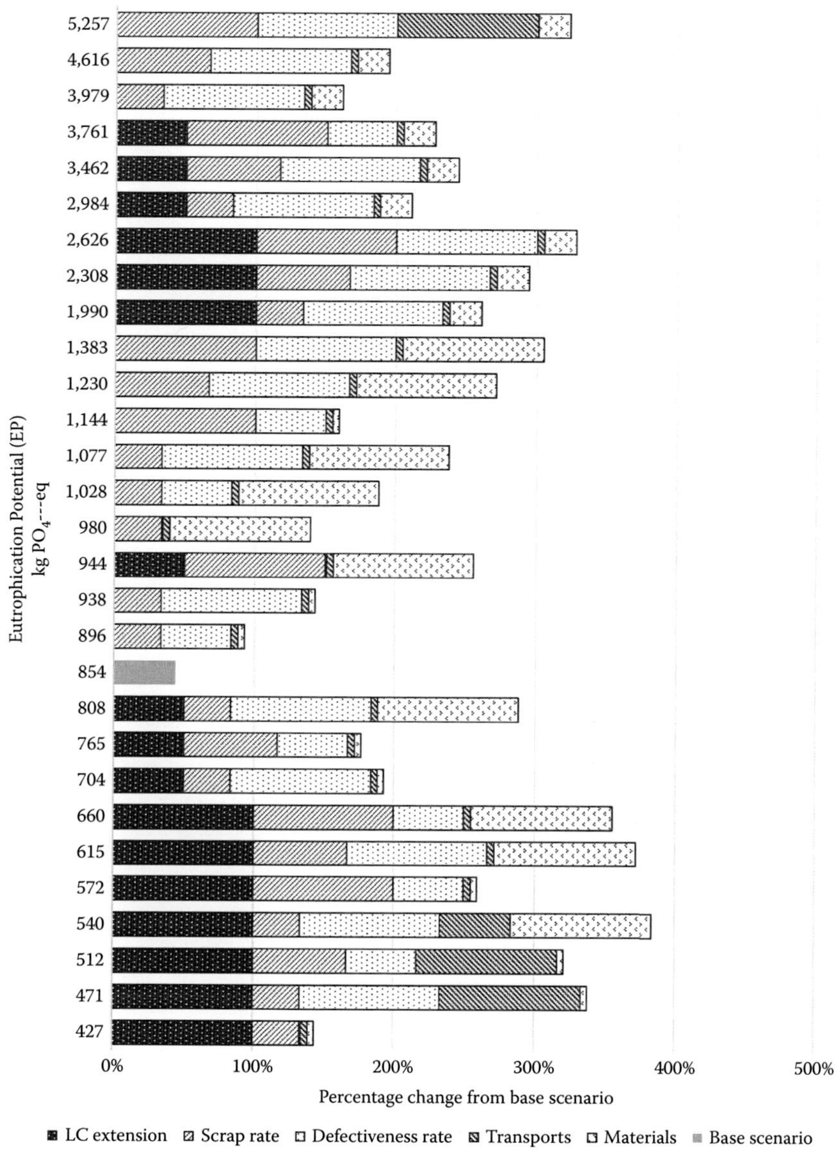

Appendix 11.4 Ozone depletion potential (ODP)

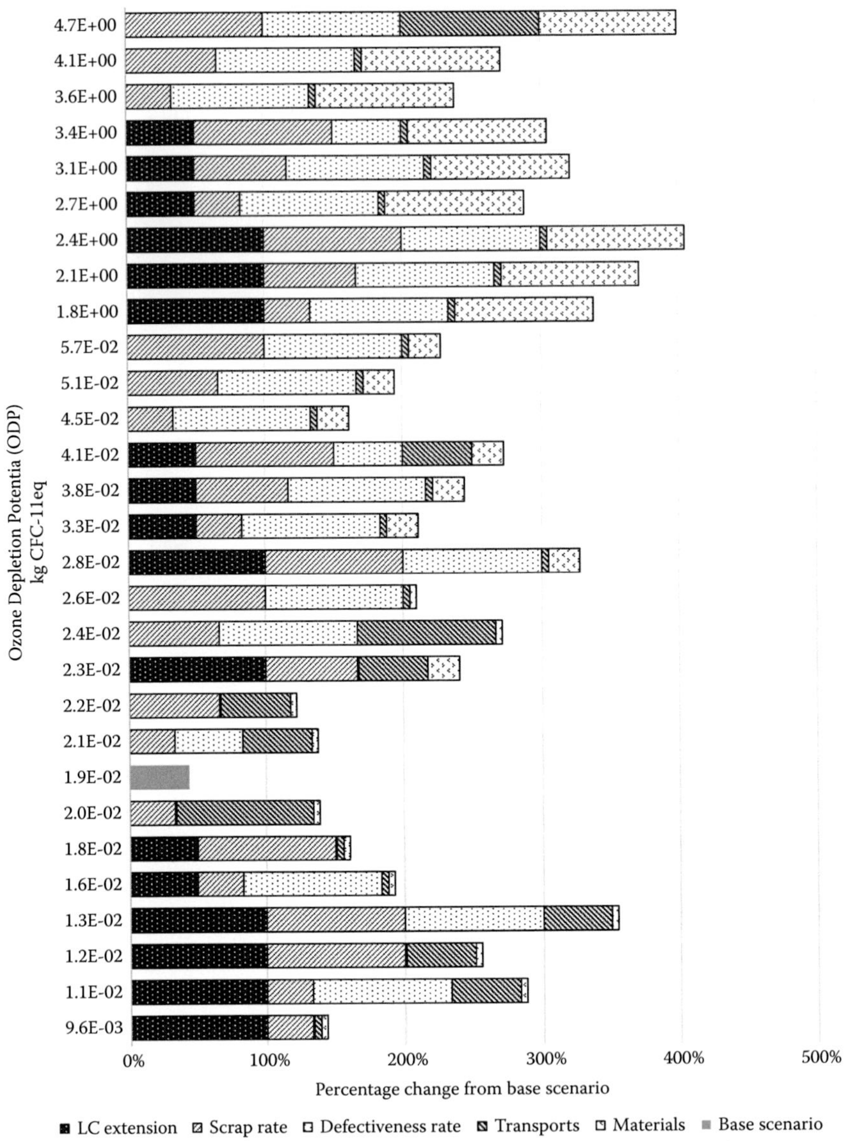

Appendix 11.5 Photochemical ozone creation potential (POCP)

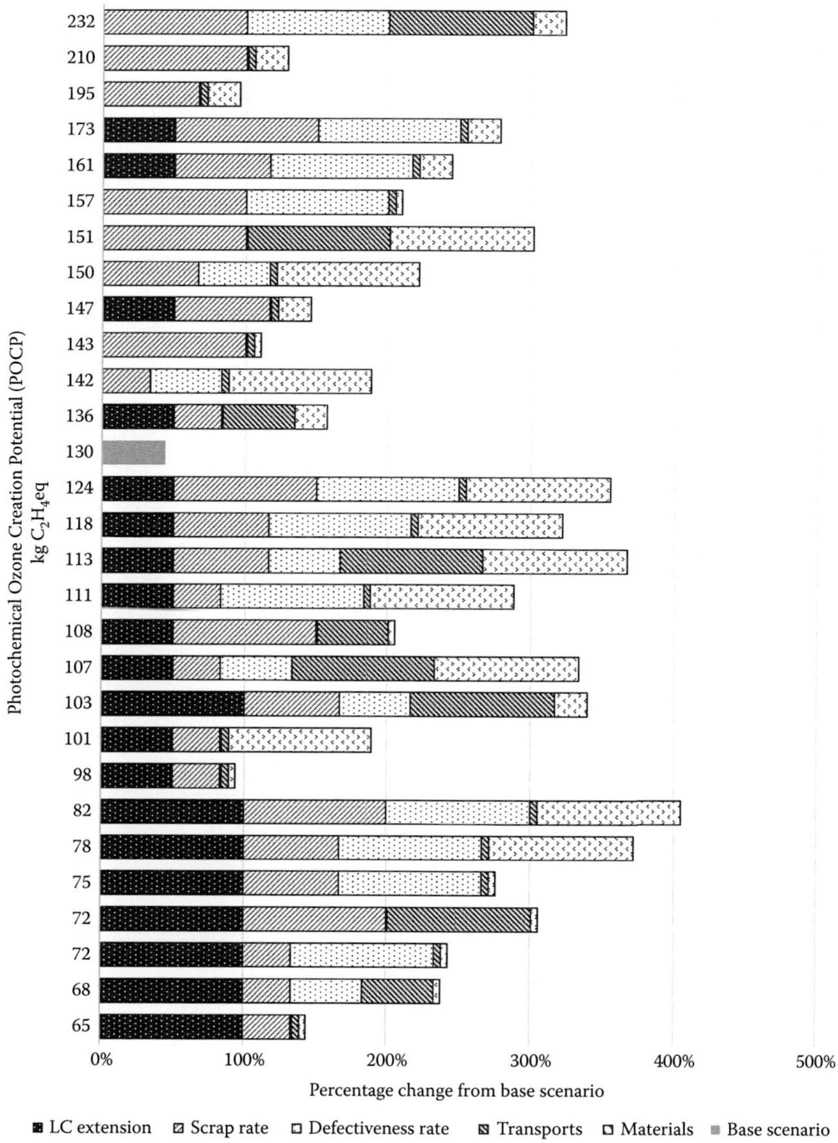

chapter twelve

Sustainability issues in mass customized manufacturing

Arun N. Nambiar

Contents

ABSTRACT

Stiff global competition and volatile customer demand have become the mainstay of today's industrial and service sectors. With more and more countries opening up their economies, companies are striving hard to stay competitive and are constantly seeking new venues that provide them with the necessary advantage over their competitors. The mass customization paradigm provides companies with the opportunity to create a niche for themselves by providing customized solutions to customer needs. Environmentally conscious customers combined with increased federal and state regulations have forced companies to focus on the environmental impact of their operations while trying to maximize the value of their products and services to keep their costs down. Sustainable development combined with mass customization affords new synergies that can potentially accrue benefits in the form of new product or service designs, efficient supply chains and increased profits. Companies can leverage information systems and other new technology to obtain maximum value from their resources, thus lowering their environmental impact while enabling them to stay competitive.

12.1 Introduction

Mass customization is a manufacturing paradigm that aims to provide customized products at mass production prices. Though this concept was originally introduced in the 1960s, it has been gaining significant traction in recent years due to stiff market competition and ever-changing customer preferences. Companies are shifting from a one-size-fits-all approach to providing highly customized products and services designed to satisfy the needs of most customers.

While this is a good idea and definitely a market winner if done right, it engenders a plethora of problems for the manufacturing and production side of businesses, not to mention the logistical challenges. Manufacturing and production facilities have traditionally been optimized to run the same set of operations over and over again in a highly efficient manner. With this shift to highly customized products, the production mix has changed from one of low variety, high volume to a high-variety, low-volume one. This creates a new optimization scenario where the parts and production operations required for this high-variety product mix need to be efficiently managed to minimize waste and maximize throughput. Modern manufacturing paradigms such as quick-response manufacturing (Nambiar 2010), lean principles (Womack et al. 1999), and agile manufacturing (Nambiar 2009a), when combined with mass customization (Davis 1989; Pine 1999), purport to create a win–win scenario for companies by helping them produce highly customized products and services quickly and efficiently.

In recent years though, there has been an increased emphasis on "going green." This is essentially a moniker in the common parlance for all things related to sustainable development. Sustainable development has been defined in the Brundtland Report (UN 1987) as "the development that meets the needs of the present without compromising the ability of future generations to meet their own needs." Thus, it can be seen that the focus of sustainable development is to utilize minimal resources, produce minimal waste, and cause minimal pollution. The three main pillars of sustainable development focus on the economic, sociological, and environmental aspects of development, respectively. That is to say, the products and services resulting from a sustainable endeavor must be economically viable for both the customer and the business, sociologically acceptable, and environmentally friendly. With the advances in information technology, customers are better informed about the environmental impact of their purchases and the manufacturing practices of the products they purchase. This engenders an informed customer base that is shifting toward environmentally friendly products and shunning products and companies with a poor track record of sustainable practices. As a result, companies are hopping onto this green bandwagon both as a marketing tool to gain more customers in today's highly uncertain and volatile market that is rife with global competition and as part of their corporate social responsibility (CSR).

Marketing gimmicks aside, sustainable practices do have significant benefits. Through these practices, fewer raw materials may be used in production, and existing materials may be reused or recycled to obtain the maximum benefit. This is of particular interest since countries such as Brazil, Russia, India, China, and South Africa (often known as BRICS) are poised to grow significantly in the coming decades, thus creating an increased demand for goods and services, which in turn places undue strain on the already limited resources. This has been quite evident in this past decade, with China's voracious appetite for natural resources to sustain its manufacturing facilities and unprecedented growth, necessitating forays into many resource-rich countries, especially in the African continent. Although China's economy has been cooling in the past couple of years, which was to be expected, other parts of the world are growing as well. The combined aspirations of large swathes of populations in developing countries for lifestyles similar to those of the developed nations creates unprecedented demands on limited resources and puts undue stress on the planet through the resulting developmental activities. Hence, instituting sustainable practices both for material conservation and environmental preservation can really go a long way in ensuring a sustainable planet for future generations.

Though the general premise behind this initiative is laudable and its benefits purportedly manifold, there are significant challenges to overcome in order for sustainable practices to become the norm and to obtain maximum benefit through the collective efforts of everyone involved. The first important challenge is changing mind-sets and encouraging people to take a more holistic view, thus enabling them to see the larger impact of their actions. This is a drastic change in perspective from the "not in my backyard" or "not my problem" attitudes that tend to be more prevalent. Oftentimes, companies take a narrow approach to sustainability by focusing on the environmental impact alone, which brings to the fore the next challenge. It is imperative to focus on the economic and sociological aspects of sustainable development along with the focus on the environment. This is important because in order for products and services that are the result of sustainable practices to become popular, they have to be affordable and acceptable to the consumers. Companies also tend to focus on the proverbial low-hanging fruit, which provide immediate results and instant gratification. However, a holistic approach that focuses on the entire value chain and critically examines all operations through the sustainability lens in order to identify areas of innovation and improvement will provide greater returns on investment in the long term. Information systems can serve as a key enabler in overcoming some of these challenges by integrating various partners of the value chain into a central system. This brings to light the fourth challenge of getting these disparate systems to communicate with each other and seamlessly integrate into a single system. Researchers the world over are focusing on ways to overcome these challenges and issues involved in the successful implementation of sustainable practices.

With mass customization, the aim is to produce products and services that customers want at prices they can afford. This takes care of two aspects of sustainable development—being economically viable and sociologically acceptable from the customer's perspective. In order for a business to leverage this to gain market share and increase profit margins, it is imperative that these efforts be economically viable from the manufacturing, production, and logistics perspectives as well. Moreover, being environmentally friendly helps from a marketing standpoint. Thus, it can be seen that sustainability initiatives can be combined with mass customization efforts to reap the maximum benefits from providing highly customized products and services at affordable prices in an environmentally friendly manner.

This chapter explores the sustainability issues involved in mass customized manufacturing. The chapter identifies the key areas of concern and examines best practices to handle sustainable development in a mass customization environment.

12.2 Background

Porter's theory of competitive advantage originally proposed in the 1980s (Porter 1985) suggests that companies can stay competitive by adopting one or a combination of two strategies.

- *Product differentiation*: By providing customers with products that meet their expectations and are different from those of the competition, companies can build a reputation that will attract more customers.
- *Cost leadership*: Companies can compete by offering products at lower prices while remaining profitable through the ensuing increased sales volume.

Modern manufacturing paradigms seek to address these tenets either individually or in a combined fashion. For example, the primary objective of the lean principles is to reduce waste, which in turn reduces costs, thus providing the cost leadership advantage. Quick-response manufacturing and agile manufacturing seek to reduce the design-to-market times in an effort to provide customers with products that meet their requirements and expectations before the competition, thus providing the product differentiation advantage. Mass customization also seeks to provide companies with a competitive advantage through product differentiation. However, in today's globalized economy, it is impractical to demand a cost premium for customized products, hence companies have to strive to not only provide customized products at a lower cost but also seek out other venues to reduce their operating costs even further.

In this mad race to lower operating costs, companies are off-loading routine activities to subcontractors and vendors while focusing on their core competencies. This allows companies to direct valuable resources toward what makes them unique and different from the competition. Mass customization is a concept originally introduced by Stan Davis (1987) and later developed by Joseph B. Pine that seeks to provide companies with the much-needed competitive advantage through product differentiation and cost leadership. Pine (1999, p. 7) defines mass customization as "providing tremendous variety and individual customization, at prices comparable to standard goods and services ... with enough variety and customization that nearly everyone gets exactly what they want." The underlying premise for this paradigm is that there is high demand for customized goods and services that are tailored specifically to meet individual customer needs and expectations. There are four basic types of customization identified as the four faces of mass customization (Gilmore and Pine 1997). These vary depending on the level of involvement of the customer in the design of the product and the level of customization possible for the product. These four faces are the *adaptive, cosmetic, transparent,* and *collaborative* approaches. These focus primarily on the end product and how it is modified or used based on individual customer needs. However, with expanding world markets and stiff global competition, it becomes imperative that companies adopt a combination of these approaches rather than one or the other.

The four approaches have been modified from the original representation in Gilmore and Pine (1997) to suggest that a company may choose its position anywhere within the continuum, similar to the production continuum proposed by Lampel and Mintzberg (1996). Moreover, with increasing operations crossing geographic boundaries, distribution and delivery are also important venues for customization, ensuring customers receive what they want, where and when they want it. Companies also tend to differentiate themselves from the competition by providing support services such as installation, repair and maintenance, recycling, and replacement. Hence, customization can be achieved at eight different levels (Da Silveira et al. 2001): standardization, usage, package and distribution, services, specific customization, assembly, fabrication, and finally design. These eight levels can be correlated to the four faces mentioned earlier. For example, packaging and distribution customization could be more of a cosmetic change or design customization could be as a result of a more collaborative effort between the company and the customer. Needless to say, irrespective of the slight differences in the focus area for customization, the ultimate objective of the company is to provide customers with products and services that meet their expectations in the hopes that this will translate into increased market share. However, stiff global competition leads to price wars where companies try to outcompete each other by providing goods and services at the lowest cost, often undercutting their profit margins. This makes it crucial to invest in mechanisms and methodologies that allow companies to lower their operating costs. Another approach might be to seek differentiation through increased emphasis on the environmental impact of their operations, given that the environmental debate about global warming and greenhouse gas emissions has been at the forefront in recent years.

Resource-based theory (Hart 1995) builds on Porter's theory of competitive advantage by suggesting that a firm's core competency arises from the value being added by its resources, and hence resources are critical to the company's success. These resources include

- *Physical*: Machines, facilities, workforce
- *Financial*: Capital to be invested
- *Tacit*: Knowledge and experience acquired over time

Hart (1995) further underscores the importance of sustainable development by proposing a three-pronged natural resource–based approach toward achieving competitive advantage. This includes

- Preventing pollution by minimizing greenhouse gas emissions and effluents, thus lowering costs through increased efficiencies, lower material costs, and regulatory compliance
- Developing new products that require fewer raw materials and can be reused or remanufactured to lower the overall cost of the product throughout its life from cradle to grave
- Engaging in sustainable practices that minimize the long-term impact of the company's operations both on the environment and the society in which it operates through investment in research to develop technologies that achieve that goal

This is in keeping with the tenets of sustainable development that were so eloquently captured by Theodore Roosevelt (2016): "Our duty to the whole including the unborn generations, bids us to restrain an unprincipled present-day minority from wasting the heritage of these unborn generations." In a subsequent follow-up (Hart and Dowell 2011) to the original propositions in Hart (1995), the authors find that there is a dearth of research in the area of sustainable development and how it provides companies with the necessary competitive advantage to make it more viable. This is despite the prevalence of numerous standards such as ISO 14000 (ISO 2016b) and ISO 14040 (ISO 2016c), which purport to codify the procedures for companies to adopt in order to implement sustainable practices. These standards have not caught on as much as other ISO standards related to quality, such as ISO 9000 (ISO 2016a). However, despite the lack of a strong causal relationship (Brunoslash et al. 2013) between mass customization and sustainability, it can be seen that sustainable practices could lead to new products, improved efficiencies, lower operating costs, and increased customer loyalty. There are numerous benefits to be accrued by combining sustainable initiatives and mass customization initiatives, due to the inherent similarities.

12.3 Sustainability and mass customization

As seen in the previous discussion, companies stand to gain a lot by applying mass customization principles in concert with sustainable practices. Both these paradigms are geared toward helping organizations to ultimately lower costs and both often result in new products and processes. It has been shown (Medini et al. 2012) that demand management is one crucial area when attempting to incorporate sustainability considerations into customization. Some of the key aspects of both these paradigms are examined in the following section, with a special emphasis on the synergies to be achieved by combining both approaches.

12.3.1 Product design

The design of the product or service is the most crucial element in sustainable development and mass customization. Products have to be designed to meet the customer's needs and requirements. At the same time, designers have to take into consideration sustainability aspects such as the impact on the environment and the ability to reuse or recycle. Through an appropriate choice of materials and production processes, the tenets of sustainability can be incorporated into the design of the product. Modular design, which is critical for the success of mass customization (Kumar 2004), also helps with sustainability since modules can be reused or disassembled for remanufacture. Concepts such as *design for engineering* (Bevilacqua et al. 2007; Glavic and Lukman 2007) and *environmentally conscious quality function deployment* (Kaebernick et al. 2003) incorporate consideration for sustainable practices into the regular design process. Some of the popular design approaches include the following:

- *Design for environment* (DfE) ensures that the environmental impact of the product both while being manufactured or assembled and being used is taken into consideration during the design phase.
- *Design for manufacturability* (DfM) focuses on ensuring that the design process includes consideration for all aspects of the product including manufacturing, assembly, shipping, and service.
- *Design for disassembly* (DfD) emphasizes the ease of taking apart a product after the end of its life and reusing its parts. This is particularly relevant and has gained in popularity with the growing emphasis on sustainability.
- *Design for recyclability* (DfR) focuses on alternate uses of parts or products at the end of their life.
- *Design for reuse* aligns the design process to focus on reusing parts in their original form after their initial use has ended.

A holistic product design model (Howarth and Hadfield 2006; Osorio et al. 2014) that takes into account input from all the stakeholders in a combined approach is indispensable. These stakeholders include the customer, regulatory agencies, the community, suppliers, and production. In the mass customization framework, products are often referred to as being *codesigned* by the designer and the customer. By incorporating sustainability principles, the design is often termed *eco-design*. Thus, through a combination of codesign and eco-design, companies develop products and services that meet customer expectations in all aspects while achieving regulatory compliance and meeting their social responsibilities.

12.3.2 *Life cycle management*

Much of the attention in life cycle management has been on handling the different growth phases (Levitt 1965) of a product, such as development or introduction, growth, maturity, and finally decline. The focus is on the external influences on the product's demand growth and how the enterprise can handle the varying demands at a strategic level. However, another related aspect is the process life cycle (Hayes and Wheelwright 1979), where the focus is on the production process used to fabricate or assemble the product. As the product evolves through the aforementioned stages, the production process used also evolves from a purely job-based shop where a large variety of products are produced in low volumes to an automated assembly line where large volumes of a single family of products are assembled.

Looking at this continuum through the prism of mass customization, it can be seen that this paradigm is designed to use medium- to high-volume production processes to manufacture or assemble a large variety of products. Thus, it becomes imperative that sustainable manufacturing practices be used through the life cycle of the customized product. On one end of the spectrum, highly automated systems tend to be designed for the efficient use of resources such as raw materials, thus minimizing waste. However, these fully automated systems can be energy intensive, and hence there is a need to examine the efficiencies of these systems and explore alternative sources of energy for running them. At the other end of the spectrum, processes are not as automated and subsequently not as energy intensive. However, these systems are not designed to be efficient in terms of their use of raw materials, primarily due to the economies of scale. Hence, there is a need to apply lean principles (Womack and Jones 2003) to continuously improve the operations in an effort to minimize waste. Mathematical models have been developed (Hu and Bidanda 2009) to take into account these various factors influencing the design of the product throughout its life cycle. Irrespective of the production process chosen, flexibility (Nielson and Brunoslash 2013a) is an essential capability in order to be able to quickly respond to changes in demand. A holistic approach to life cycle management that takes into consideration the impact of product design, production processes, and product usage, and subsequent recycling or reuse helps companies extract the maximum value from its products.

12.3.3 Sustainable production

Production is an important aspect of any manufacturing enterprise and it assumes all the more importance in a sustainable environment since it is often an energy-intensive process too. There have been different production approaches to implementing mass customization. The most successful is the modular approach, where the features of the product are separated into modules and can be quickly plugged together in different ways to provide a customized product. This approach requires the standardization of parts across multiple products or product families, thus providing the economies of scale required to produce these individual components in large volumes. Concepts from lean principles such as value stream mapping, continuous improvement, and pull systems can be employed to help make the production process more sustainable. Some of the benefits of applying lean principles to the production process include the following:

- *Value identification*: Though the concept of the value chain has been around for quite some time (Porter 1985), the popularity of lean principles, which underlie the Toyota Production System, has brought renewed focus on value as perceived by the customer and the stages through the product chain where value is created or added, also known as the *value chain*. Value stream mapping (Womack and Jones 2003) is a technique within the lean principles that helps companies identify the value of the product for its customers and examine the activities that contribute to that value. Since a lot of the success of mass customization hinges on the company's ability to provide more value to its customers than the typical mass-produced products, knowledge about the value of its products and its value chain can be capitalized to improve its product offerings. This also helps with sustainable development, since a focus on value and creating value would automatically result in an efficient use of the available resources.
- *Waste reduction*: As per lean principles (Womack and Jones 2003), every activity within the enterprise can be classified into three broad categories: value-added activities, necessary non-value-added activities, and unnecessary non-value-added activities. Clearly, by definition, the most desirable is the value-added group of activities. By critically examining the entire value stream in detail, the non-value-added activities can be identified and eliminated.
- *Inventory minimization*: Eliminating wastes and redundancies in the production process allows the process to function just in time with minimal if not zero inventory. This is particularly important in a mass customization framework because of the inherent vagaries in the product demand. With modular design and just-in-time assembly, the inventory of finished goods can be kept to the bare minimum. This allows companies to respond to changing demands in an agile manner (Nambiar 2009b).
- *Order-driven production*: One of the important tenets of lean principles is the *kanban* system or *pull* system. This is essentially an order-driven process where downstream operations "pull" parts or products from their upstream counterparts, who in turn pull from their upstream operations. This ensures that there is no unnecessary inventory build-up. This is again of utmost importance in a mass customization environment since companies would not want to be saddled with an inventory of customized products with no demand.

12.3.4 Sustainable packaging

One of the eight generic levels of mass customization identified by Da Silveira et al. (2001) is packaging. This correlates to the *cosmetic* face among the four faces of mass customization as put forth by Gilmore and Pine (1997). In this form of customization, the focus is on how the product is packaged for each segment of customers. In many cases, the product inside is the same irrespective of the external packaging. For example, candies may be wrapped in orange and black during Halloween or in red and green during Christmas. The product may also be available in different quantities. For example, the same breakfast cereal may be available in to-go containers for one-time use or in family-size value boxes. This is one of the easily achieved forms of customization since there is very little change in the product and is hence more widely adopted. However, with the growing emphasis on sustainability, companies are looking toward minimizing the use of materials for packaging their products. This focus on sustainable packaging has multiple advantages, such as

- *Reduced cost*: In many cases, examining the packaging can help identify better and cheaper alternatives that are more sustainable, thus lowering costs.
- *Faster shipping*: As a result of efficient packaging, more products can be fit into a standard shipping container, thus facilitating more products to be shipped at once and at the same price.

Companies have already begun finding innovative ways to implement sustainable packaging. For example, Amazon's (2016) frustration-free packaging ships many products in their original box, thus saving on costs while providing customers with a pleasant experience opening packages. Apple is purported to have a separate packaging room (Lashinsky 2012) where researchers spend countless hours trying to get the packaging right. Some of the mechanisms of implementing sustainable packaging include

- *Minimalist packaging*: Minimalist design can be achieved by reducing the packaging to the bare minimum necessary to ensure that the product arrives in good condition, and it provides customers with a good experience unraveling the product. This ensures that very little raw material is used to begin with.
- *Recycled material*: Another approach is to use recycled material in product packaging. This may be through the use of recycled paper or plastic.
- *Recyclable*: A third aspect is ensuring that all the packaging material is recyclable. Some companies go beyond simply ensuring that the material is recyclable by offering to collect the packaging material to be recycled.

Companies use one or more of these approaches in an effort to ensure minimum impact on the environment while providing customized products to meet customer demand.

12.3.5 Sustainable supply chain management

Improving customer satisfaction and streamlining material flow rank high among the critical factors motivating companies toward efficient supply chain management (Tummala et al. 2007). In a mass customization environment, the problem of material flow is compounded many times over due to the sheer variety of products. Thus, efficient supply chain management is crucial for the success of mass customization initiatives. Due to increasing pressures from environment-conscious customers and regulating bodies, combined with the ensuing competitive advantage, companies are exploring ways to make their supply chain more sustainable. This necessitates a closed-loop supply chain (Nielson and Brunoslash 2013b) that takes into account sustainable issues such as energy consumption and waste along with other aspects of a typical supply chain such as cost, time, and quality.

Sustainable or green supply chain management has been defined (Srivastava 2007, pp. 54–55) as "integrating environmental thinking into supply chain management including product design, material sourcing and selection, manufacturing processes, delivery of the final product to the consumers as well as end-of-life management of the product after its useful life." An integral part of green supply chain management is the reverse supply chain, which includes five main activities (Guide and van Wassenhove 2002) geared toward the reverse flow of material.

- *Acquiring* the product through an efficient collection system spread across all supply chain partners
- *Transporting* the collected products to a central facility
- *Inspecting* and *sorting* the products and their constituent parts
- *Reusing* and/or remanufacturing parts and finally
- *Distributing* recycled or remanufactured products

The large number of products in a mass customized industry has adverse effects (Shui-Mu and Su 2013) on the reverse supply chain since it makes the aforementioned steps in the process all the more arduous. However, in order to stay competitive in today's market, characterized by stiff competition and shrinking profit margins, it is imperative that companies strive to recover as much value from their products once they reach the end of their lives. In order to achieve this, it is imperative to create a strong corporate culture of cooperation and collaboration among supply chain partners (Trappey and Wognum 2012; Liu 2013). It is also important to have the necessary support system to achieve this collaboration in the form of a robust information system and an efficient communication network. There is also an increased emphasis on the need for the supply chain to be more agile to be able to respond to uncertainties (Tachizawa and Thomsen 2007) in product demand in terms of variety and quantity and the uncertainties in deliveries given the global nature of the market.

12.3.6 Sustainable logistics

The Council of Supply Chain Management Professionals (Ballou 2007, p. 338) defines logistics as "that part of supply chain management that plans, implements, and controls the efficient forward and reverse flow and storage of goods, services, and related information between the point of origin and point of consumption in order to meet customer requirements." Logistics assumes increased importance in a mass customization environment because of the sheer complexity of the operations with increased product variety. Hence, an efficient logistics system is indispensable (Gooley 1998) in a mass customization framework. It is also important for the logistics system to be flexible (Hamid 2015) in today's truly globalized and highly volatile markets. The system should be able to quickly respond to the uncertainties inherent in a mass customization framework. This response also needs to be in an efficient manner so that companies can still stay competitive. This underscores the need for sustainable logistics with increased focus on the use of "clean" vehicles with zero or low emissions. This is also especially true because transportation accounts for a significant percentage of the greenhouse gas emissions released into the atmosphere.

Another area of growing importance is reverse logistics. Reverse logistics has been defined (Rogers and Tibben-Lembke 1999, p. 2) as "the process of planning, implementing and controlling the efficient cost-effective flow of raw materials, in-process inventory, finished goods, and related information from the point of consumption to the point of origin for the purpose of recapturing or creating value, or for proper disposal." This is an integral part of the reverse supply chain that is essential for the sustainable use of limited resources. Reverse logistics typically involves a network of players each performing a vital function of the process. The vital functions in reverse logistics (Srivastava 2015) are

- *Collection*: Once a product has been expended, it is important to urge consumers to return the product to the manufacturer in order to maximize the value of the product. This process of returning goods to the manufacturer needs to be easy and economical in order to motivate consumers. Many establishments have begun having collection points at various locations for consumers to return their no-longer-needed products. The geographic location of these collection centers is critical in order to make them effective.

- *Inspection*: The collected products have to be inspected to evaluate their current state. This is important because consumers often return products in various stages of disarray, which determines the recovery process.
- *Sorting*: The products are then sorted into various categories based on material recoverable, ease of recovery, and so on.
- *Processing*: The products often undergo some sort of processing to recover parts or raw materials to be recycled or reused. In some cases, it might be as simple as taking apart the product and cleaning up the parts for reuse, while in other cases the parts may have to be processed into raw materials for alternative uses.

Supporting these various functions is the underlying logistical support and distribution network that is so critical to the successful implementation of reverse logistics. It can be seen that there are a lot of common features between the forward supply chain and reverse supply chain, the primary being the distribution network. One important difference, though, is that oftentimes speed is not an issue in reverse logistics, whereas it is of utmost importance in the forward supply chain. Despite the critical difference, there are numerous advantages to be accrued through the synergy of forward and reverse supply chain networks. For example, as the trucks make deliveries in the forward supply chain, they can also be collecting recyclable and reusable products from the various collection centers. Reverse logistics is particularly important in the electronics industry, especially with the rapid proliferation of mobile phones and other computing devices. It has been shown (Srivastava 2015) that implementing an efficient reverse logistics mechanism and in turn a reverse supply chain system can provide a competitive advantage through increased return on investment and improved customer image.

12.3.7 *Information systems*

The successful implementation of sustainable and mass customization practices requires an integrated information system (Mahajan et al. 2002; Frutos and Borenstein 2004; Dean et al. 2008; Ngniatedema 2012) that connects all the value chain partners while providing an effective communication network for sharing information (Bai and Gu 2010). It can be seen that supply chain integration is crucial for mass customization due to the sheer diversity of products and the volatility of demand.

In order to be able to quickly respond to changing customer demands, it is imperative to be able to communicate the customer needs through the entire value chain. This becomes even more critical with numerous parent companies adopting drop shipping, where customer orders are often fulfilled directly by the supplier. With increased emphasis on the reverse supply chain as a result of the growing impetus for recycling and remanufacturing parts, it is essential that supply chain partners work closely with each other. Information systems serve as the key enabler in providing the infrastructural support needed for increased collaboration and cooperation between value chain partners. There are two approaches to building the necessary information systems infrastructure:

- *Centralized*: A centralized networked system with real-time monitoring will allow for informed and instantaneous decision-making regarding product mixes, production volumes, and product distribution and collection.
- *Decentralized*: In this scenario, each player in the value chain implements their own information system with the caveat that these systems should be able to communicate and interact with other systems in the value chain. This can be accomplished by adopting commonly accepted design patterns and ensuring uniform standards for data encoding.

In either case, it is essential that the information system is capable of being scaled up and down to reflect the growth patterns of the organization. It is also important for the system to be flexible and easily adaptable to the changing market conditions. Ensuring data security and integrity is also an essential feature of the information system. The challenges involved in building such an information system scaffolding include

- *Diverse needs*: The requirements of individual players are so diverse and disparate that it is difficult for a single unified system to address all of the needs.
- *Legacy systems*: Many organization already have an existing technology infrastructure that they have been using for many years and continue to use because all their data conforms to that system. It is often difficult for these legacy systems to interact with newer systems without intermediate data exchange mechanisms.
- *Piecemeal implementations*: For many years, information systems and technology was more of an afterthought, and hence piecemeal implementations have been put in place by organizations to address specific needs as they arise, thus making seamless integration an onerous task.

- *Multitude platforms and devices*: The proliferation of mobile devices of all forms and sizes combined with the burgeoning range of platforms catalyzes into cyclopean combinations, which engenders compatibility issues.

12.4 Future research direction

Even though sustainability and mass customization concepts have been around for more than a few decades, there is a heightened sense of urgency and need to focus on these aspects in today's world for numerous reasons such as stiff global competition and stringent environmental regulations, to name a few. Both these concepts have significant overlap in their reliance on robust information systems, agile supply chains, and efficient product design. In this chapter, some of the benefits and challenges involved have been identified. There needs to be more research into the benefits of combining the two concepts based on actual evidence. Though concepts such as design for excellence (DfX) and its offshoots, such as DfE and DfM, have been applied in isolation for specific objectives such as improving production efficiency or reducing environmental impact, there needs to be greater integration between these systems to facilitate efficient and effective product design. Reverse logistics is a vital and significant component of sustainable supply chain management, and the synergies afforded by combining forward and reverse logistics need to be explored and exploited. Since a lot of the success of these principles relies on information sharing and exchange, it is important to have well-defined industry-wide data formats that allow individual, disparate systems to communicate with each other across the multitude of platforms and provide a seamless user experience across the various devices, ranging from handhelds to desktops. As with any new concept or paradigm, change management is a significant issue, especially if it requires a complete overhaul of the existing operations. In order to reduce some of the implementation throes, it might also be useful to better communicate and share some of the best practices from companies that have been successful, since sustainability is about the collective good for future generations.

12.5 Conclusion

Globalization has provided customers the world over with access to a wide variety of products while providing enterprises with access to new sources of raw materials and skilled workforces. This has made customers more demanding, thus engendering stiff competition among companies to gain market share. Mass customization is gaining popularity in this milieu since it purports to provide customized products and services at mass production prices. However, in this mad rush to drive down costs in order to stay competitive, many enterprises have become highly distributed, with operations in many parts of the world, and highly diversified, with a plethora of products. This has created a supply chain nightmare. With global warming and drastic climate change around the world, there is increased attention on the environmental impact of human development. Sustainable development has assumed paramount importance and companies are striving hard to reduce the environmental footprint of their operations. A synergy of sustainable development and mass customization could lead to efficient product or service design, agile and flexible supply chains, and robust information systems, thus providing companies with the much-needed competitive advantage to survive in this era of globalization.

References

Amazon. (2016). About Amazon certified frustration-free packaging. Retrieved from https://www.amazon.com/gp/help/customer/display.html?nodeId=200285450. Accessed on January 7, 2016.

Bai, J., and Gu, C. (2010). Research on model of information sharing in mass customization supply chain. *Proceedings of the 2010 International Conference of Logistics Engineering and Management: Logistics for Sustained Economic Development: Infrastructure, Information, Integration,* Chengdu, China. pp. 4221–4226. American Society of Civil Engineers.

Ballou, R.H. (2007). The evolution and future of logistics and supply chain management. *European Business Review.* 19(4). 332–348.

Bevilacqua, M., Ciarapica, F.E., and Giacchetta, G. (2007). Development of a sustainable product lifecycle in manufacturing firms: A case study. *International Journal of Production Research.* 45(18–19). 4073–4098.

Brunoslash, T.D., Nielsen, K., Taps, S.B., and Joslashrgensen, K.A. (2013). Sustainability evaluation of mass customization. *Proceedings of the International Conference on Advances in Production Management Systems,* State College, PA. Vol. 414, pp. 175–182.

Da Silveira, G., Borenstein, D., and Fogliatto, F. (2001). Mass customization: Literature review and research directions. *International Journal of Production Economics.* 72(1). 1–13.

Davis, S. (1989). From future perfect: Mass customizing. *Planning Review.* 17(2). 16–21.

Dean, P.R., Tu, Y.L., and Xue, D. (2008). A framework for generating product production information for mass customization. *International Journal of Advanced Manufacturing Technology.* 38(11–12). 1244–1259.

Frutos, J.D., and Borenstein, D. (2004). A framework to support customer-company interaction in mass customization environments. *Computers in Industry.* 54. 115–135.

Gilmore, J.H., and Pine, B.J. (1997). The four faces of mass customization. *Harvard Business Review.* 75(1). 91–91.

Glavic, P., and Lukman, R. (2007). Review of sustainability terms and their definitions. *Journal of Cleaner Production.* 15. 1875–1885.

Gooley, T. (1998). Mass customization: How logistics makes it happen. *Computers and Industrial Engineering*. 37(4). 49–54.

Guide, V.D.R., and van Wassenhove, L.N. (2002). The reverse supply chain. *Harvard Business Review*. 18(2). 25–26.

Hamid, J. (2015). Logistics flexibility: A systematic review. *International Journal of Productivity and Performance Management*. 64(7). 947–970.

Hart, S.L. (1995). A natural resource-based view of the firm. *Academy of Management Review*. 20(4). 986–1014.

Hart, S.L., and Dowell, G. (2011). A natural-resource-based view of the firm: Fifteen years after. *Journal of Management*. 37(5). 1464–1469.

Hayes, R.H., and Wheelwright, S.C. (1979). Linking manufacturing process and product life cycles. *Harvard Business Review*. 57. 133–140.

Howarth, G., and Hadfield, M. (2006). A sustainable product design model. *Materials and Design*. 27. 1128–1133.

Hu, G., and Bidanda, B. (2009). Modeling sustainable product lifecycle decision support systems. *International Journal of Production Economics*. 122. 366–375.

ISO. (2016a). ISO 9000: Quality management systems. Retrieved from http://www.iso.org/iso/home/store/catalogue_tc/catalogue_detail.htm?csnumber=45481. Accessed on January 8, 2016.

ISO. (2016b). ISO 14000: Environmental management. Retrieved from http://www.iso.org/iso/home/standards/management-standards/iso14000.htm. Accessed on January 3, 2016.

ISO. (2016c). ISO 14040:2006: Environmental management; Life cycle assessment; Principles and framework. Retrieved from http://www.iso.org/iso/catalogue_detail?csnumber=37456. Accessed on January 3, 2016.

Kaebernick, H., Kara, S., and Sun, M. (2003). Sustainable product development and manufacturing by considering environmental requirements. *Robotics and Computer Integrated Manufacturing*. 19. 461–468.

Kumar, A. (2004). Mass customization: Metrics and modularity. *International Journal of Flexible Manufacturing Systems*. 16(4). 287–311.

Lampel, J., and Mintzberg, H. (1996). Customizing customization. *Sloan Management Review*. 38(1). 21–30.

Lashinksy, A. (2012). *Inside Apple: How America's Most Admired and Secretive Company Really Works*. John Murray, London.

Levitt, T. (1965). Exploit the product life cycle. *Harvard Business Review*. 43. 81–94.

Liu, L. (2013). Supply chain management under mass customization. *Advanced Materials Research*. 616–618. 2044–2047.

Mahajan, V., Srinivasan, R., and Wind, J. (2002). The dot.com retail failures of 2000: Were there any winners? *Journal of the Academy of Marketing Science*. 30(4). 474–486.

Medini, K., Da Cunha, C., and Bernard, A. (2012). Sustainable mass customized enterprise: Key concepts, enablers and assessment techniques. *Proceedings of the 14th IFAC Symposium on Information Control Problems in Manufacturing, Bucharest, Romania*. Vol. 14, pp. 522–527.

Nambiar, A.N. (2009a). Mass customization: Where do we go from here? *Proceedings of the 2009 World Congress on Engineering*, London, UK, July 1–3. Vol. 1, pp. 687–693.

Nambiar, A.N. (2009b). Agile manufacturing: A taxonomic framework for research. *Proceedings of the 2009 International Conferences on Computers and Industrial Engineering*, Troyes, France, July 6–9. pp. 684–689.

Nambiar, A.N. (2010). Modern manufacturing paradigms: A comparison. *Proceedings of the 2010 International Multiconference of Engineers and Computer Scientists*, Hong Kong, March 17–19. Vol. 3, p. 1662–1667.

Nielsen, K., and Brunoslash, T.D. (2013a). Assessment of process robustness for mass customization. *Proceedings of the International Conference on Advances in Production Management Systems*, State College, PA. Vol. 414, pp. 191–198.

Nielsen, K., and Brunoslash, T.D. (2013b). Closed loop supply chains for sustainable mass customization. *Proceedings of the International Conference on Advances in Production Management Systems*, State College, PA. Vol. 414, pp. 425–432.

Ngniatedema, T. (2012). A mass customization information systems architecture framework. *Journal of Computer Information Systems*. 52(3). 60–70.

Osorio, J., Romero, D., Betancur, M., and Molina, A. (2014). Design for sustainable mass-customization: Design guidelines for sustainable mass-customized products. *Proceedings of the International Conference on Engineering, Technology, and Innovation: Engineering Responsible Innovation in Products and Services*, Bergamo, Italy. IEEE Explore, pp. 1–9.

Pine, B.J. (1999). *Mass Customization: The New Frontier in Business Competition*. Harvard Business School Press, Boston, MA.

Porter, M.E. (1985). *Competitive Advantage*. New York: Free Press.

Rogers, D.S., and Tibben-Lembke, R.S. (1999). *Going Backwards: Reverse Logistics Trends and Practices*. RLEC Press, Pittsburgh, PA.

Roosevelt, T. (2016). Theodore Roosevelt Association. Retrieved from http://www.theodoreroosevelt.org. Accessed on January 9, 2016.

Shui-Mu, H., and Su, J.C.P. (2013). Impact of product proliferation on the reverse supply chain. *Omega*. 41. 626–639.

Srivastava, S.K. (2007). Green supply chain management: A state-of-the-art literature review. *International Journal of Management Reviews*. 9(1). 53–80.

Srivastava, S.K. (2015). Network design for reverse logistics. *Omega*. 36(4). 535–548.

Tachizawa, E.M., and Thomsen, C.G. (2007). Drivers and sources of supply flexibility: An exploratory study. *International Journal of Operations and Production Management*. 27(10). 1115–1136.

Trappey, A.J.C., and Wognum, P.M. (2012). Network and supply chain system integration for mass customization and sustainable behavior. *Advanced Engineering Informatics*. 26(1). 3–4.

Tummala, V.M.R., Phillips, C.L.M., and Johnson, M. (2007). Assessing supply chain management success factors: A case study. *Supply Chain Management: An International Journal*. 11(2). 179–192.

Womack, J.P., and Jones, D.T. (2003). *Lean Thinking: Banish Waste and Create Wealth in Your Corporation*. Simon & Schuster, New York.

Womack, J.P., Jones, D.T., and Roos, D. (1999). *The Machine that Changed the World*. Rawson Associates, New York.

UN. (1987). Report on the World Commission on Environment and Development: Our common future. Retrieved from http://www.un-documents.net/our-common-future.pdf. Accessed on January 3, 2016.

Name Index

N

O

P

Q

R

S

T

Subject Index

D

N

Q

R